'A unique and groundbreaking contribution to innovation through bio-inspired design. One of the most inspiring books of the last decades, which profoundly boosts eco-effective innovations to grasp desperately needed disruptive changes for a planet with 10 billion people.'

Professor Michael Braungart, co-author of *The Upcycle* and *Cradle to Cradle*

'[A] richly detailed, meticulous, well-written book. These well-crafted tales of bio-inspired innovation will entrance general readers and warrant the close attention of scientists and technologists.'

Kirkus starred review

'Instead of trying to crudely dominate the world around us, it's nice to learn that more and more smart humans are trying to figure out how we might use the clues from other species to fit in a little more easily on this tired old planet.'

Bill McKibben, bestselling author of *Eaarth*

'A skilled journalist and science writer, Khan makes complex topics easy to understand as we travel around the world to meet scientists, engineers, and the plants and animals that are inspiring breakthrough solutions to some of our greatest technological challenges. She doesn't shy away from in-depth explanations of the biology and the engineering behind innovations from material science to nanotechnology to robotics and more, while bringing researchers' stories of discovery to life through her enjoyable, informative writing style. This book is a worthwhile read on many levels.'

Jay Harman, author of *The Shark's Paintbrush: Biomimicry and How Nature is Inspiring Innovation* and CEO of PAX Scientific

About the Author

Amina Khan is a science writer at the *Los Angeles Times.* She's covered the Curiosity's landing on Mars and explored abandoned gold mines in pursuit of a dark matter detector. This is her first book.

AMINA KHAN

ADAPT

How We Can Learn From Nature's Strangest Inventions

Atlantic Books
London

First published in the United States of America in 2017
by St. Martin's Press.

First published in trade paperback in Great Britain in 2017
by Atlantic Books, an imprint of Atlantic Books Ltd.

This paperback edition published in 2018

10 9 8 7 6 5 4 3 2 1

A CIP catalogue record for this book is available
from the British Library.

Paperback ISBN: 978 1 78649 229 6
E-book ISBN: 978 1 78649 228 9

Printed in Denmark by Norhaven

Atlantic Books
An imprint of Atlantic Books Ltd
Ormond House
26–27 Boswell Street
London
WC1N 3JZ

www.atlantic-books.co.uk

To Anwar and Nayyar Khan

CONTENTS

Part IV: SUSTAINABILITY

PROLOGUE

In Douglas Adams's five-book trilogy The Hitchhiker's Guide to the Galaxy, dolphins are the second-most intelligent species on Earth, right after mice. Humans come in third, an intellectual also-ran to their furry and finned superiors.

The story of hapless anti-hero Arthur Dent managed to tap into that uncomfortable, itchy malaise humans sometimes get: a suspicion that we're not as smart as we think we are. We generally try not to make eye contact with this feeling, for fear that it might just look back.

However, avoiding the truth comes with its perils. In the Hitchhiker universe, humans were so incompetent that they ignored the dolphins' frantic warnings that the world was about to end, dismissing their hoop-jumping and tail-twirling as a particularly acrobatic aquatic show. Then the Earth was destroyed to make way for an interstellar superhighway.

Clearly there's one thing the science-fiction classic got dead wrong. When it comes to the intelligence of the species, humans probably rank much lower on the list.

I developed this uncomfortable feeling while chaperoning grade-schoolers at the California Science Center in Los Angeles. As a science writer at the *Los Angeles Times,* I fully expected to impress the kids. In front of the space shuttle *Endeavour,* I told saucer-eyed students tales of meeting the ship's astronauts. I feigned nonchalance while one dipped her fingers toward a starfish into the still, chilly waters of the "touch tank." But then a strange shape under the ripples drew my eye. There, amid the jewel-toned starfish and the lazy sea urchins, was a strange, purple-black corkscrew big enough to fit perfectly in the palm of my hand.

"That's a shark egg," the volunteer behind the pool told my eight-year-old charge.

I perked up and leaned over, nearly elbowing my little girl out of the way. Amid the pebbles sat the strangest, largest screw I'd ever touched. Its smooth, tapered threads spiraled down to a point, the encasing hard but flexible, like fingernails after an hour in the bath.

I'd never seen anything like it. From the fish to the ostrich, animals lay round eggs. Their smooth curves distribute force and minimize breakage. They shouldn't come in squares, or triangles, or other pointy shapes. But this design—the same one scattered around my dad's workbench—allowed the shark to wedge her unhatched babies into a rock, where they'd be more difficult for predators to remove. It's an engineering design that's been in use for millions of years. Just not by people.

It's humbling. Humans have a tendency to think we're at the top of the creative pyramid—the brains amid the beauty, brawn and the plain-old-bizarre creatures that inhabit the earth. Everything amazing is something we made up out of whole cloth.

But we're behind the curve in so many ways. Nature isn't just our equal. With a four-billion-year head start, it has surpassed human ingenuity beyond our wildest imaginations.

This stark disparity dawned on me a few years back at a conference on fluid dynamics. What seemed like a dull topic turned out to be anything but dry. Thousands of researchers converged on Long Beach to explain the aerodynamics of flying snakes and the secrets of sharp-toothed shark skin.

Amid talks of hummingbird helicopters and ants' crawling patterns, I learned that scientists are increasingly turning to the biological world for inspiration and education—trying to understand the apparently miraculous in order to discover new engineering secrets. This is not entirely new. Observant humans have certainly lifted a trick or two from nature's book in the past. George de Mestral created the ubiquitous Velcro after pulling several tenacious burrs off his dog's fur and then, curious about their super-sticking properties, discovered their loopy hooks under a microscope. The revolutionary warping wing designed by the bird-watching Wright brothers allowed their plane to safely turn like the earth's natural aviators.

Yet these have registered as brief, isolated blips in a culture that, since the Industrial Age cranked into high gear in the nineteenth century, has largely viewed nature as something to be tamed, fixed, overridden, ignored, and even destroyed. We have solved most technical problems through making things bigger and more energy and resource intensive. For better or for worse, it has gotten us results. But a few centuries of heedless consumption have left us near ecological and manufacturing dead-ends. We're approaching certain limits of engineering. There is no more low-hanging fruit. The problems left to be solved—in medicine, in architecture, in computing—are complex and intractable. Plus, we're running out of raw materials, poisoning the environment. The brute force methods that got us so far are now failing us.

So researchers, at least, have started paying attention to how nature succeeds where we have not. Biologists have begun to realize that their explorations of the natural world apply to other realms. Meanwhile, engineers have begun to notice that biologists may hold answers to many of the most unsolvable questions in physics. It's a mode of thinking that's picked up major momentum in the last few decades, and it has a name: biologically inspired design.

In this book, we'll meet scientists from very different disciplines coming together to learn from biology and push the limits of our engineering abilities. We'll go from the very small scale (the chemistry of photosynthesis) to very, very big (the principles of ecosystems). It will be divided into sections based on four themes: material science, mechanics of movement, architecture of systems, and sustainability. Each chapter will explore a discipline where new discoveries have been made, and more appear on the horizon, as people examine how nature has outperformed our current technologies. Over the course of this book, we'll look at all of these examples and more to explore how adapting nature's innovations to improve human technology will allow us to do more things not bigger, but better.

I will follow scientists in the lab and the field as they conduct their breakthrough research. I'll peer under the microscope with scientists studying the nanoscale properties of cuttlefish skin. I'll head out to the San Gabriel Valley where engineers with the Defense Advanced Research Projects Agency are testing the next generation of humanoid robots. I'll head to Namibia with a group of biologists and engineers who each have their unique reasons to study the six-foot-tall termite mounds dotting the savannah.

The concept of bioinspired design first gained traction under the term "biomimicry." A book by the same name, written by Janine Benyus in 1997, crystallized this cross-disciplinary idea for many researchers. According to a 2010 report commissioned by San Diego Zoo Global, in fifteen years biomimicry could make up $300 billion annually of the United States' GDP—and $1 trillion of the world's output—in 2010 dollars. It could also mitigate the depletion of various natural resources and reduce carbon dioxide pollution, putting another $50 million back in our collective pocket. "Biomimicry could be a major economic game changer," the authors wrote. Its commercial use "could transform large slices of various industries in coming years and ultimately impact all segments of the economy."

We now realize that most new findings will not come from blindly imitating nature wholesale, but from studying it to learn its most invisible secrets in illuminating detail. Many of the natural world's mysteries play out in realms where scientists' understanding of physics is fuzzy at best and perilously off-base at worst. Without studying nature's secrets, says Geoffrey Spedding, an engineering professor at the University of Southern California, "You can miss a phenomenon completely, get it completely wrong, not just a little bit wrong."

By paying close attention, researchers are gaining remarkable insights along the way. Gecko feet stick to walls without any need for adhesive, harnessing the weakly interacting van der Waals forces. Snakes can fly simply by rearranging their ropy bodies. Common bean leaves can stop bedbugs in their tracks—without any need for pesticides—by using a vicious array of stabbing hooks that have thus far proven impossible to replicate with synthetic materials.

Weird and wonderful as these discoveries seem, they're increasingly vital in a world where we're running out of resources, in which we need to learn to live sustainably, using fewer harsh chemicals and creating less waste. The first step is to learn how other livings things have been doing so, with great success, for billions of years.

Some researchers are already working in revolutionary ways—breaking down the barriers between biology and engineering to find out what they can learn from one another. It's cutting-edge work, and it's producing remarkable, potentially world-altering results.

In the years since that mind-altering physics conference, I've written about how scientists are learning from humpback whales' knobby fins to make better wind turbines, and studying the jellyfish as a model for the human heart. There are researchers learning from ant colonies to control traffic and studying organisms to design better cities.

All of this requires researchers to think beyond the confines of their own discipline, and to connect with others outside of their own field. It also applies to many different scales, from nanotechnology to city planning, and it affects countless areas of research and application, from medicine to architecture. Because of this vast span of disciplines and scales, it has been a challenge to find guiding principles that researchers and innovators can follow to seek out and apply bioinspired solutions.

Efficiency is a powerful driver of nature's many forms and functions, the high virtue of bioinspired engineering. Some of evolution's most astounding innovations occur because it's dealing with limited resources, or trying to survive harsh environments, or repurposing an already-existing biological quirk for a totally new function. (That's how birds first took flight—their feathers were once little more than dinosaur decoration and insulation.)

That's why biology seems to know the secrets of fluid dynamics better than we do and why it appears to be such a skilled architect at nanoscales. These and other areas of nature's expertise will continue to pop up throughout this book.

If necessity really is the mother of invention, the mother of all inventors is Mother Nature. And while nature didn't come up with a wheel, it can build a pretty decent screw. The trick, bioinspired enthusiasts say, is to take the strategies seen in nature and learn from them—maybe even improve on them, too.

It's remarkable how much you can pick up from the natural world if you pay attention. I've reread The Hitchhiker's Guide more times than I can count, and I surf, so the ocean's charms sometimes seem routine to me. But I learned recently that I hadn't taken those lessons about the intelligence of dolphins to heart.

One morning in Florida, as other surfers and I struggled past the pumping whitewater, I saw two dolphins hanging out just before the break. I figured we were simply sitting in their fish, and paid them no mind. Then a wave loomed—one that none of the shortboarders dared take.

The dolphins lined up to face the beach and tilted their noses downward as the peak picked them up and carried them forward. My mouth fell open. They were surfing—like the oldest of old-school pros. If dolphins had ten, they'd have been hanging them.

A third dolphin leaped over the pair as they took off, just to drive the point home.

Perhaps they were trying to send a message. Though I'm pretty sure it was less a warning of the apocalypse and something more along the lines of, "We've seen seaweed that surfs better than you amateurs."

But if we pay attention to what dolphins, and birds, and the rest of nature is telling us, we may be able to find a way to save it—and ourselves—before it's too late.

PART I

MATERIALS

1

FOOLING THE MIND'S EYE

What Soldiers and Fashion Designers Can Learn from the Cuttlefish

An air-shredding volley of bullets headed straight toward your soft, woefully underarmored body can have a powerful clarifying effect on your most recent life choices. *What's happening? Where are they shooting from? How can I hide?*

Once might be bad timing, an unlucky brush with death. But when the bullets keep coming, on different days and in different places, the question changes: *Why does this keep happening?* You start to look a little further back for answers. Bad luck starts to look more like a bad pattern.

A bad pattern was exactly what kept getting soldiers in the U.S. Army nearly killed, according to Major Kevin "Kit" Parker. Parker is a professor of bioengineering and applied physics at Harvard University, but two decades ago he was just a Southern boy who'd decided to join the army partway through graduate school, completing basic training in 1992 and getting commissioned as an officer in 1994.

"Military service is a little bit more common in my family or in the neck of the woods where I'm from—so, you know, if you watch NASCAR and you're very susceptible to good advertising, you might find yourself in the army," Parker says, with a laugh.

After joining the Army Reserves, Parker ended up serving two tours of duty in Afghanistan, from 2002 to 2003 and in 2009, and twice in 2011 as part of a special science advisory mission called the gray team. The 2009 tour was particularly rough, a seven-month stretch when Parker's unit just couldn't seem to duck the militants. Wherever they went, their convoys kept getting pinned down by gunfire.

"It was a very rough combat tour; I was getting shot at quite a bit," Parker said. "One day I was out with some Afghan national police and we were on this kind of desert plain that was on the other side of a mountain. There was no vegetation, nothing—and I'm looking at my shirt, this . . . bluish-green pixelated pattern, and I'm looking at the dirt around me and I thought, 'I stick out like a sore thumb here!' "

The problem was the camouflage on their uniforms. Known as Universal Camouflage Pattern, or UCP, it was rolled out in 2004 to the tune of $5 billion after several years of development. Blue, green, and pixelated, the design was meant to be an all-terrain garment that would eliminate the need for multiple uniforms. But instead of letting them blend in to all environments, this one-shade-fits-all suit made the army major and his fellow soldiers stand out against what was often a barren, rocky landscape.

"This was a budget-driven decision, rather than a science-driven decision," Parker said.

There was a combat cameraman on the day that Parker looked around his blue-suited body and had his horrible realization. The cameraman took a photo of Parker down on one knee—an image that would serve as inspiration when he arrived back home.

"All I had to do was kind of look back at my photographs

from the war and I see that picture of me out on one knee out in the desert," Parker said. "It's like slightly less conspicuous than if I'd been holding a big sign over my head in Pashtun that said, 'Shoot me.' "

Parker wasn't the only one with this problem. The camouflage was making soldiers in Afghanistan easy targets—and in 2009 the issue finally reached the ears of now-deceased U.S. Rep. John Murtha (D, Pennsylvania), who reportedly heard from noncommissioned officer Rangers while on a visit to Fort Benning, Georgia. Study after study began to come out showing that UCP was a sub-par camouflage. One report in particular, conducted by U.S. Army Natick Soldier Research, Development and Engineering Center, showed that four other camouflage patterns performed 16 percent to 36 percent better than UCP across the test's woodland, desert, and urban settings. At least one of those, known as MultiCam, had been available since 2002—which means a $5 billion expense on the research and development of these uniforms could have been avoided.

According to news reports, the issue wasn't just the colors, or the pixelation (a technique used by several more successful camouflage patterns). The problem was also the pattern's scale. It was too small, and suffered from a phenomenon known as "isoluminance," where a pattern's colors are so close together that they blend when seen from a distance and make the entire form stand out. In the case of UCP's light-toned colors, the soldiers' outlines became light-colored silhouettes, making them easy to see against the background. In other words, it could be making the soldiers more visible and thus, less safe.

In the wake of public and insider outcry over UCP, MultiCam was adopted as a temporary fill-in pattern for Afghanistan; a new pattern similar to MultiCam was reportedly debuted in 2015. But when it comes to developing new, effective military camouflage, Parker said, "We still aren't getting it right."

There has to be a smarter way to approach camouflage than coming up with a one-size-fits-all pattern, Parker thought. The problem hung in the back of his mind after he returned from his second tour of duty. And then, in the fall of 2009, some two months after returning home, Parker got a call from Evelyn Hu, an optical physicist at Harvard, inviting him to work on a project funded by the Defense Advanced Research Projects Agency—DARPA, that defense department outfit that midwifed the Internet in the 1960s and 1970s and still funds cutting-edge, futuristic research today. Hu's project, however, was at least fifty million years old: the cuttlefish, an alien-looking sea creature that—at least in the United States—is less well-known than its close relatives the octopus and the squid.

While the cuttlefish may not be as recognized as its eight-armed cousins, it rivals them in a number of aspects, including its intelligence and its incredible shape-shifting, shade-shifting skin. The animal can change its coloration in about 300 milliseconds. Hu wanted to partner up with a marine biologist named Roger Hanlon, a researcher at the Marine Biological Laboratory (MBL) in Woods Hole, Massachusetts, and a leading expert in cephalopod behavior (the group that includes cuttlefish, squid, octopuses, and nautiluses). And Parker realized that Hanlon's office just happened to be nearby.

"I said, you know what? I'm right here. I'm going to walk over there and get this guy," Parker said. The three scientists teamed up with several other colleagues and went on to publish a 2014 paper on the nanoscale color-changing mechanisms within cuttlefish skin.

Parker happened to be visiting the Marine Biological Laboratory's library when he got Hu's call—a library near whose doorway hangs the somewhat ironic inscription, STUDY NATURE, NOT BOOKS.

It was said by Louis Agassiz, a groundbreaking biologist who helped inspire the creation of the MBL. It's one of Hanlon's favorite quotes. He cites it often, in joking defense of his research habits.

"Certainly that's my excuse to get out in the world and go diving a lot," Hanlon says.

That's something of an understatement. Over his roughly thirty-five-year career, the biologist has performed around five thousand scuba dives, everywhere from Australia to South Africa and the Caribbean. (Among his favorite spots: Palau Islands of Micronesia and Little Cayman Island.) But back in the lab, he has a whole slew of captive cuttlefish whose stealth and smarts he and his colleagues can study on a daily basis.

I'm sitting in his office in Woods Hole, Massachusetts, at the Marine Biological Laboratory, one of the research stations hanging along the cape, just down the street from the dock that takes summer vacationers to the island known as Martha's Vineyard. It's a cold, clear, and quiet day in December, and the light has a thin, golden touch as it hits the enclosed harbor known as Eel Pond. Hanlon's office looks out onto the water, a tiger-striped *Nautilus pompilius* shell on the windowsill punctuating the view. Books on his shelves have a certain theme: *Vision and Art: The Biology of Seeing*; *Neurotechnology*; *Butterflies;* and a beast of a book with the no-nonsense title, *Disruptive Pattern Material* stamped across its double-wide spine.

"An interesting book," Hanlon notes, when I point it out. "That guy made a fortune off of clothing."

Hanlon is a sort of Renaissance man of marine camouflage, but for the most part he studies all manner of cephalopods—various species of squid, octopus, and cuttlefish. When I ask him which his favorite is, he laughs, almost surprised. "The European cuttlefish is pretty phenomenal," he says. "The one we have in our lab here. I've worked a lot on them; it's a really neat animal."

In the United States, the cuttlefish has long been the lesser-known cousin of the octopus. They inhabit the coasts off of Europe, Asia, Australia, and Africa, but somehow seem to skirt the Americas. Like their squid cousins, they have eight arms and two tentacles. Aristotle admired their iridescent innards; during his time, the animals were prized for their ink, which they spew out just as the octopus and squid do in order to throw up a defensive curtain and escape. They've long been called the "chameleons of the sea," for their ability to blend in to their surroundings.

"These animals also escape detection by a very extraordinary, chameleonlike power of changing their colour," Charles Darwin wrote in his seminal 1860 book *The Voyage of the Beagle.*

You may not be able to think of a weirder, more otherworldly creature than the cuttlefish. It has no backbone, strange W-shaped pupils in its bulbous eyes and thick floppy-looking arms protruding from its face. It swims around with a tutu-like frill that floats around its body, but propels itself backward to escape predators. To look at its many-fingered face is to look into the visage of Cthulhu, that fictional god of H. P. Lovecraft's strange horror stories, or the alien Ood of the rebooted cult TV show *Doctor Who.*

If you think there's little similarity between a human and a tuna fish, consider this: at least they both have backbones. As a member of the cephalopods, the cuttlefish is from an even more distant branch of the family tree than true fishes. Cephalopods, an extremely mobile group of animals that includes fellow shade-shifters like the octopus and the squid, are thought to be the brainiest group of invertebrates on the planet.

The modern cephalopod lineage first emerged more than five hundred million years ago, before even sharks had come to be. They arose from animals known as mollusks—a group whose living members include the humble snail and the clam, which is essentially a muscle armored with a shell.

How did the very clever cuttlefish evolve from an extended family that even today includes such, well, *brainless* creatures? The answer may lie in the shell—or the lack thereof, as it were. Mollusks were defined by their protective, calcium-rich armor—the word "mollusk" comes from the Latin *molluscus,* meaning "thin-shelled." But while a shell can act as highly effective defense for vulnerable flesh, it can also be a burden—and at some point in their evolution, cephalopods abandoned their external shells, becoming highly mobile hunters and foragers in the ocean. Unlike the squid and the octopus, the cuttlefish does have a flat and oval internal shell called a cuttlebone, which is full of layered chambers. The front chambers are filled with gas and the rear chambers are filled with seawater; and by adjusting the ratio of gas to liquid, the cuttlefish can change the density of the cuttlebone—and thus, its floatability—as it swims around at different depths. That's a big energy saver if you're trying to move around and stay afloat. (Cuttlefish also save energy by living a quiet life, often lying on the seafloor and using their arms to toss sand over their bodies so that they're hidden from sight.)

The trade-off to not having a shell is that you become a very tempting target for every other predator in the ocean. You're basically a fluid bag of meat. A prepackaged, protein-filled snack. In short: you are pretty easy pickings.

Octopuses and squid, cephalopods who have also shed their shells, suffer from the same vulnerability. So these fleshy creatures have evolved an ingenious system of defense: if you can't protect yourself when attacked, don't draw any unwanted attention in the first place. Swim under the proverbial radar. And so they've developed this highly specialized camouflage that seems able, at first glance, to match nearly any color under the sea, if not the sun. It's not a totally unique ability—animals like chameleons can also change color, depending on their mood. But few can do it with the sophistication of the cephalopods, which can not only change color but change intricate patterns that allow them to quickly blend into their surroundings, whether it's the sandy seafloor or a bundle of wavy kelp. They can even modify the texture of their skin to match their environment, whether it's rough sand or a sharp-edged reef.

I get a firsthand view of this when Hanlon takes me down the hall to the lab, where the cuttlefish reside. His colleague Kendra Buresch is there in a room that sounds like an Escher sketch of running faucets—tanks full of floating cuttlefish line the far wall, and water circulates through at a high rate, as it would in an ocean environment. The animals are smaller than I expected—roughly the size of my hand—but just as cute as I thought they'd be. (It doesn't hurt that, to my ear, "cuttlefish" sounds like "cuddlefish.")

These guys aren't interested in cuddling, though. As Buresch comes close, one of the animals lifts two of its arms up in the air—almost as if begging for food (though I'm told later by Hanlon that it's a startle or threat response). They definitely have personalities, Buresch said, as she held a finger over the water. The cuttlefish focuses on her finger, and two dark, slightly wavy lines grow and thicken across the length of his back; they remind me of the twin flowing f-holes carved into a violin. The longer the cuttlefish stares at Buresch's wiggling finger, the thicker those markings get—as if someone's trying to fill in a narrow line with a blunt, wet sharpie, and the color is beginning to bleed outside the lines.

"So now he has a pattern on, that's frequently the pattern they get when they're hunting," Buresch says. "I just don't want him to grab me—it's fine, I just don't like the way it FEELS!"

Buresch's voice rockets up two octaves at that final word, as the cuttlefish, who can stand it no more, lashes out with its two feeding tentacles and latches onto her finger. Buresch pulls back quickly and the cuttlefish releases its suckered grip, the dark lines on its back fading from front to back and quickly flickering away.

I totally want to try now. The once-fooled cuttlefish won't be tricked again. Buresch waves her finger in front of him to make sure and he thinks about it—the lines start to draw themselves up his back, but just as quickly vanish. (They're not in synchrony, so the effect looks somewhat like an alternating eyebrow-waggle, Stephen Colbert–style.)

So I wiggle my finger in front of his companion, who seems a little more trigger-happy. The dark lines appear, and the little guy shoots out his tentacles, hidden behind his eight-armed face, soft little suckers wrapping around my finger. It's weird but not unpleasant, and I'm thinking about how to describe the feeling and kind of entranced by how the tentacles are shortening, making it look like the cuttlefish is reeling itself in toward my hand, and then I notice that Buresch and her colleague, Stephen Senft (who walked in about a minute before this) are making noises of significant alarm.

I pull my finger back and he doesn't let go; I pull up, out of the water, expecting him to release—but at this point, his tentacles are so short that he just hangs on for dear life, his single-digit meal almost in reach. So I inadvertently lift him into the air. The shock (or perhaps the gravity) is finally too much for him, and he lets go and plops back down into the shallow tank.

The researchers are noticeably relieved and a little shocked. Senft says he's played with the cuttlefish before, but has never gotten as close to having his finger become a snack as I did.

"They have a sharp little beak in there," Senft says by way of understatement. (Its cousin, the squid, has a beak that's measured to be among the hardest, if not the hardest, all-organic material known—which I'll get into in the next chapter.) Meanwhile, I'm feeling pretty bad about messing with this poor cuttlefish. I ask the scientists if I traumatized the little guy.

"He'll be okay," Buresch says.

Animals that rely on camouflage often make use of pigment molecules that absorb most wavelengths of light and reflect only a tiny slice of the visible spectrum. This works well for most animals with yellows and reds and brown/blacks in their coats (particularly mammals, whose pigments are limited by what their hair can produce). The iridescent shades and blue greens like those you'll see in the blazing sapphire of a peacock feather are thanks to structural color—building a nanoscale surface that does not absorb light but reflects incoming wavelengths into the blue-green range.

Some animals, like chameleons, can actively change colors at will. Those that can't alter their appearance on command instead feature large-scale patterns—alarming false eyespots, eye-fooling tiger stripes—that, while static, are remarkably good at disrupting the eye and keeping the wearer out of harm's way.

The cuttlefish takes advantage of all of these colors and tactics. It can produce red, yellow, and brown; it can create bluey greens and even whites—but it also deploys them the way that permanently camouflaged animals do, using clever patterns to fool the brains of both predator and prey.

I see one of these patterns in action as Buresch turns toward a blue tarp tent—an ice-fishing hut, as it turns out—and unzips it. Inside is a kiddie pool–sized clear plastic tank, in which, inside a large pizza-sized black tub, a pocket-sized cuttlefish sits next to a little checkerboard-covered pillbox. Above the pool-in-a-pool, a camera stares directly onto the spot where the cuttlefish sits.

As Buresch slowly reaches in to grab the black-and-white

box—painstakingly slowly, so as not to alarm the cuttlefish—the little animal's pattern starts to change. A dark spot flares up onto the left side of its back—the side closest to Buresch's approaching hand—and another one begins to emerge on the right. They're similar to the dramatic "eyespots" you'll see on butterfly wings and the backs of fish, meant to startle or confuse an oncoming predator into thinking it's looking into the face of a different animal.

"He's trying to fool me and make himself look scary," she says.

Buresch retreats, checkered tchotchke in hand, and the slightly cockeyed spots fade entirely away. She zips up the blue tent; she'll be able to monitor him on the video screen placed next to the fishing hut. She's waiting to see if the cuttlefish changes patterns, now that a key visual cue—the checkerboard cylinder—has been removed, and all that remains is sandy substrate.

A cuttlefish's environment is visually cluttered, Hanlon had pointed out: sharp rocks, floating algae, branching coral, arm-waving sea anemones, speedy schools of fish, crawling crabs—in the water, there's a mind-boggling diversity of shape and movement. It's an overwhelming amount of information to take in, and there's no way to match it all. It's like a visual "cocktail party effect": Imagine yourself at some social soiree, trying to listen to all of the conversations around you at the same time. You'd go mad. Instead, you just try to pick out the most relevant conversation you hear and hold onto that thread, ignoring the cacophony around you.

Cuttlefish do that in their cluttered environments. They focus on a particular visual cue and then sync their camouflage with that cue. And the researchers want to figure out which cues they respond to most strongly—which ones the cuttlefish thinks are most important. That's why, in the blue ice hut, the little checkerboard pillbox had been placed in the sand next to the cuttlefish.

On the lab counter a few feet away are a few large-pizza-sized circles with different patterns on them, which I flip through—one in the same checkerboard pattern and scale as the pillbox, another in plain gray. These artificial "substrates" get slipped onto the floor of the little black tank, so that the scientists can see whether the animals respond to the ground beneath them or the significant objects (like the checkerboard pillbox) jutting out of them.

The laminated, waterproof substrates also come in smaller and larger-scale checkerboards, because the researchers want to see at what scale the cuttlefish decides it's time to turn on the bright white square it sports in the center of its back. Fanned out a few inches away are three other random-looking pixelated patterns, at different levels of contrast. They remind me a little bit of the "digital camouflage" that's the center of the army camo controversy.

Buresch also has substrates that look like plain old pictures of sand. As it turns out, the cuttlefish don't really care about the texture of the surface they're on; the ground can be flat as a board and they'll still respond to it. (I find this all the more remarkable, given the incredible texture-producing properties of their skin, but more on that later.)

Cuttlefish don't just respond to patterns—they'll try to imitate the objects around them, Buresch said, recalling an ingenious little experiment Hanlon and a few other colleagues performed. The scientists placed a single cuttlefish in a teardrop-shaped tank with a sprig of artificial seaweed placed near the rounded wall. The cuttlefish would swim over to the fake greenery and raise their arms with a jaunty little crook near the ends to mimic the branching algae. This has been reported anecdotally in the wild: divers have spotted both squid and cuttlefish waving their arms in the ocean currents to emulate the undulating algae around them, making it even harder to distinguish the animals from the background.

Cute as this artful interpretation of marine shrubbery may be, there's a deeper geometric principle behind it. The scientists next ran a stripped-down version of their experiment, placing patterns on the teardrop walls to see how they'd react. In all cases, the floor was a plain gray, but the walls were lined with thick black-and-white stripes. Sometimes the stripes were vertical, sometimes they were horizontal, and sometimes they were diagonal. And the scientists found that the cuttlefish would actually try to follow the direction of these lines with their arms, holding them straight out for the horizontal pattern, straight up for the vertical pattern, and at a slant for the diagonal patterns.

But even though the animals respond to simple, two-dimensional designs, they can turn their skin into exquisite three-dimensional structures to match their surroundings. The cuttlefish, like the octopus, can also make its skin extend these little branches, which can sprout tinier branches until their whole bodies look like a spiny fragment of coral. It's truly remarkable, Hanlon said, and still something of a mystery. They know it functions as a muscular hydrostat—the way your tongue does, with one end anchored and the other end free-moving—but it's as if your tongue was able to reform with two branching mini-tongues at the end.

That's not what Buresch is looking at in her work, however. After switching out the surface pattern beneath the cuttlefish in the ice hut, she's been giving the animal a little time to settle down before checking to see if he's adjusted the markings on his body. With no pillbox nearby, Buresch expects that he's gone from high-contrast to sandy-tone, though she has to check to be sure.

You might think, at this point, that the cuttlefish is such a master of disguise because it's able to perfectly match its surroundings. In that case, the perfect camouflage would be, essentially, the perfect invisibility cloak; it wouldn't make its wearer transparent, but would precisely match its surroundings, changing with the scenery. That certainly sounds like a tempting solution for man-made camouflage—it would solve the military's problems with UCP, enabling soldiers to adapt to tan deserts and lush forests without ever stripping down and suiting back up. But think about that idea for a moment, and it quickly starts to fall apart. First, it would require an enormous computer to process the visual environment and spit it out onto the clothing, which would be expensive to build and impractical to carry; and it would really only work if you move slowly or stand still. After all, if you're in front of a tree and then step a few inches to one side, unless your clothing works instantaneously and from all angles, you may suddenly look like a glaring brown target to a sniper scanning for your mostly defenseless body.

No two environments are exactly alike—and in fact, no single environment is exactly like itself, when viewed from different vantage points. Lucky for the cuttlefish, camouflage is never really just about looking like your surroundings (though it helps). Consider a tiger's stripes. The black stripes don't just help the animal blend in with the shadows; they also serve to break up the tiger's silhouette, because animals, humans included, are primed to look for full-body outlines. That's why a tiger's stripes suddenly cut off near the ankle—they make it harder to quickly identify.

The cuttlefish has to fool a range of predators (large fish, diving birds, sharks, and seals) and prey (fish, crabs, mollusks—a variety of small and tasty creatures). They can see at night; some can see ultraviolet wavelengths of light. Luckily for the biologists who want to understand it and the engineers who want to mimic it, the cuttlefish is not actually trying to match its environment to a T. In fact, Hanlon says, its true power comes not from its ability to fool the eye, but its talent for fooling the brain.

Here's what Hanlon started to notice, after observing countless cephalopods, cuttlefish included, react to their surroundings: In spite of the incredible detail the animals put into their disguises, each one was not unique. In fact, the vast array of camouflage patterns all seemed to boil down to three (maybe four) basic templates. So in the thousands upon thousands of slightly different situations a cuttlefish might find itself in its one-to-two-year life span, it only ever uses three different disguises. How's that for universal camouflage patterns?

That may sound pretty bonkers at first, though it starts to make a little more sense when you know what the patterns are. The first is a uniform/stipple pattern; it's low-contrast and meant to blend in with a nice, even sandy bottom. The second is a mottled template; it's higher contrast and more uneven (think light-and-dark marble-sized pebbles). The third is what the researchers call "disruptive": a very high-contrast and very large-scale pattern, a big, boxy, bold design whose main feature is a giant white square on the cuttlefish's back. This pattern is used when there's a high contrast environment (preferably with large white rocks) where there's no way the cuttlefish can really blend all the edges of its body to its surroundings. Instead, in a more dramatic (and perhaps effective) way than the tiger stripes, the cuttlefish makes faux edges inside its own body—which dramatically break up the cuttlefish's actual silhouette, making it very difficult, visually, for predator or prey to piece its body outline together.

If you don't believe that the brain can be so easily fooled, consider the Rubin vase. You've seen it before: that black-and-white image that either looks like a white chalice, or like the silhouetted profiles of two men facing each other. Both are part of the image that your eyes see, but your brain has to decide if the edge between black and white belongs to the vase, or to the men. Stare at it for several moments, and you'll feel your brain switching between the two. Vase! No, men! No, vase! And so on, until you get a headache like I just did while testing it out. The brain has to decide which line to follow—and it has a very hard time holding both images in mind at the same time. The cuttlefish's disruptive pattern, while its dramatic coloring might seem to attract attention, takes advantage of our brain's search for outlines. It creates false borders, forcing the brain to divvy up an object incorrectly, and thus miss the profiles of the animal entirely.

In an ideal world, that's exactly what smart man-made camouflage is meant to do—and what UCP was supposed to do (and did not, according to Natick-led reports in 2009 and 2010). To me, it seems the pattern didn't take into account lessons learned by many different animals, including the cuttlefish. For the cuttlefish, scale plays a key role: The pattern has to be big enough, the pieces of the puzzle large enough, that the brain doesn't see the whole for the parts. This was part of the issue with the army camouflage: the pattern was too small-scale to break up the soldiers' silhouettes. Personally, I'm starting to think that if they'd studied the cuttlefish, they might have realized that was not the best idea.

Regardless of any military experts' interest in the topic, scientists continue to study natural adaptive camouflage simply for the sake of getting the basic science down. For example, Hanlon and his colleagues are still trying to clearly demonstrate that there are only three (or four) patterns—and that takes a ton of work, he said. You have to gather thousands upon thousands of images, develop an algorithm to analyze them, and see if they all fit the pattern or if there are any outliers. It's a huge, data-driven effort, and it's difficult to prove.

But they're hacking away at the data, because it is a stunning idea. It means that, no matter the type of animal, from mammals to fish and birds, all visual predators have brains that operate by the same fundamental rules—and they can all be tricked in some basic ways. That is an idea that's not just of interest to military experts, but to any and all biologists looking to understand the behavior of different species, the environmental factors that shape their evolution, and the common cognitive structures they share.

But the researchers have to do more than simply prove that these underlying pattern templates exist. They want to understand when a cuttlefish decides to use one or the other. After all, a cuttlefish's environment can be overwhelmingly complex; it might include a nice patch of smooth sand, some mottled coral, or a bunch of large rocks where putting on a disruptive pattern would work nicely. Which one does it choose? Understanding this big-brained animal's decision-making process will shed light on what the camouflage priorities should be in a given environment.

For example, cuttlefish seem to prefer coordinating with three-dimensional objects in their environment. Put them on a sandy substrate with a light-colored rock off to the side, and they'll swim straight over to the rock and light up the white box on their back—even though they could just as well put on their sand-tone uniform pattern and hang out anywhere else in the tank.

Of course, if the rock were also sandy-looking, then they *would* pop on the uniform pattern; if it looked mottled, they'd throw on the mottled pattern. For the cuttlefish, the order of priority seems to be matching with three-dimensional objects, then two-dimensional vertical patterns (such as seaweed in the wild or the walls of the tank in Hanlon's experiments), and finally the two-dimensional patterns of the ground beneath them.

These are the sorts of questions Buresch is exploring with experiments akin to today's very informal demo—which she went to check on before I left her lab. Before she'd removed the checkerboard pillbox from the pool, the cuttlefish had swum right up to it and donned its disruptive pattern. Now that she had pulled the pillbox out, she expected (as she'd seen many times before) that the disruptive pattern would fade away in favor of a sandy, uniform tone. But when we looked at the monitor, its disruptive pattern seemed even more high-contrast than before.

"They don't like to cooperate when people are staring," Buresch says. Somehow, I'm not surprised.

Understanding why a cuttlefish chooses a particular camouflage is one thing. Understanding exactly how it manages to do this is another problem entirely. It requires a whole different set of tools, because the secret of the cuttlefish's color-changing powers lies on the scale of individual cells. Once you understand the micro-scale mechanisms that underpin rapid adaptive camouflage, you have a fighting chance to actually replicate it. Then, once you replicate these mechanisms, you can potentially mass-manufacture textiles that harness them. That end goal is what draws people like Kit Parker, who sees the promise from a military standpoint.

Before that can happen—and whether or not that happens—biologists need to understand the tiny organs, called "chromatophores," that make this system work. That's up to researchers like Stephen Senft, who's showing me a baby squid under a microscope. The squid is tiny, about the size of a very small ant; but stuck on a slide, stained in a special dye and placed in the scope, the pigment organs show up clearly as dark polkadots underneath its translucent skin.

Cephalopod skin has proven devilishly difficult to study, and it has taken several decades of painstaking examination and key developments in microscope technology to finally get scientists to a point where it's even possible to grasp the nanoscale mechanisms at work.

There are half a dozen microscopes in this building, which sits across from the building where Hanlon, Buresch, and Senft work. Before this little field trip, Senft had arrived just in time to see the cuttlefish almost take my finger off and had led me away before I could traumatize any more animals, up the stairs to the lab bench where he does much of his work. Like I said, understanding the skin's powers requires a different set of tools—mainly different kinds of microscopes: some traditional, some that use lasers, some that harness electrons in order to see structures that are smaller than the wavelength of light. And in his office, he pulls up some of the fruits of that study: strangely beautiful images of balls of different sizes, strung together like a tangled pearl necklace.

Senft is one of the scientists who worked with Parker and Hanlon and Harvard optical physicist Evelyn Hu to try and understand the complex physics of cuttlefish skin. And complex it is, made of several layers, shot through with muscles and nerves to control the colors that appear and fade on the surface, as well as the texture it takes on. And it can do this in a matter of 300 milliseconds. In animals like the chameleon, the brain sends out messages using hormones that float through the bloodstream and eventually change shades over several seconds or minutes. The cuttlefish operates much faster, because it can send messages from the brain via nerve impulses. They're incredibly long, superhighway nerves, Hanlon said—single strands running straight from the brain to the muscle, rather than being relayed along several nerves at synaptic crossings. These nerves connect to the muscle fibers controlling the chromatic organs in each layer of cuttlefish skin.

The scientists wanted to know what was happening at the

receiving end of those neural instructions: how were the cuttlefish creating yellow-brown patterns that masked their bodies, or using neon blue-green hues to show off to females, or generating the blinding white of their eye-fooling square?

Turns out that understanding how cuttlefish skin works is a multilayered operation. In other color-changing animals, chromatophores essentially just contain pigments that act as selective color filters. But, as the scientists found when they put cephalopod skin under a microscope, that's not the case at all with cuttlefish skin. It's a far more complex—and ingenious—operation.

There's a reason the scientists refer to the chromatophores as "organs," not mere pigment-filled cells, Hanlon said. There are several different cell types involved in the operation of these structures, including muscles and nerves and pigment sacs. The pigments, like the dyes in your clothing, absorb almost all incoming light and reflect back only certain wavelengths. In the case of the cuttlefish, it has chromatophores with either yellow, red-brown or brown-black pigments, each in their own layer: yellows on top of reds on top of browns.

Beneath those three layers of pigmented chromatophores are the iridophores, which operate by a totally different optical trick. These cells take advantage of a protein called reflectin, which has some pretty dazzling properties. Reflectin doesn't absorb light; it takes incoming light and manipulates it by forcing the waves to bounce off of its surface at different angles. Those crisscrossing light waves interfere with each other, creating what's known as structural color. It's the same principle behind the iridescent blues and greens you'll see inside an abalone shell, or on a butterfly wing, or even on a glistening slab of meat at the butcher shop. Pigments, like the yellow, red, and brown of the chromatophores in the upper skin layers, absorb and reflect the same specific wavelengths at pretty much every angle. But in the iridophores, platelets composed of reflectin shoot the wavelengths out at different angles all over its surface, which is why, if you move an iridescent object from side to side, changing your viewing angle, it seems to shift colors. It's a beautiful mastery of optical physics—and a dazzling sight to behold.

Cuttlefish have certainly learned how to wield their varied color palette well. While the primary purpose of their color-changing skin may be keeping the animals hidden, out of sight, they have also learned to harness it for hypnotizing their prey. A cuttlefish will put on a show for an unwary crab, running colors across its skin like it's a Las Vegas marquee. The hapless crab, unable to look away, will just sit there, apparently transfixed. Still flickering, the cuttlefish will inch ever closer until—bam!—its two tentacles shoot out, grab the crab, and pull it toward its mouth. Imminent death never looked so pretty.

And now we've reached the final, lowest layer of cuttlefish skin responsible for color: the layer of leucophores. Leucophores have an essential task: they produce the whites in the cuttlefish skin. And while white may seem like a fairly, well, vanilla color compared to the psychedelic, shade-shifting iridophores, it's one of the most challenging layers to understand—and it's the layer that Stephen Senft focuses on.

Like the iridophores, the leucophores also carry granules made of that strangely folded protein, reflectin. But in the iridophores, those reflectin proteins are incorporated into little plate-shaped structures; in the leucophores, they're built into spheres. So instead of producing the shiny, shifting hues that the iridophores do, the leucophores produce one of the whitest whites in both the natural or man-made world. What does whitest white mean? It means whiter than milk. It means whiter than paper. Senft, sitting at a laboratory bench one floor above Hanlon's office, pulls up a graph showing the spectrum of light of paper and of the cuttlefish leucophore. Keep in mind, white light is made up of all the colors of the visible spectrum, from red to indigo. But the graph showing the paper's whiteness is full of sharp dips where certain wavelengths—that is, certain colors—are missing. The graph for the leucophores, on the other hand, is essentially flat and even all the way through—few to no dips. It includes nearly all wavelengths equally, and is thus a nearly perfect white.

Senft in particular is interested in understanding how it is the leucophores do what they do. He's committed to cracking open the black box that these white-producing sacks of protein represent. Among the mysteries: why some of the granules are spherical, and others are flat as plates. Presumably, if you want to create an all-over white, your granules would all be spheres—but that's clearly not the case. Senft and his colleagues suspect that there's an optical reason for the plates' existence, as well as the direction in which they're aimed and the way they're spaced out. The cuttlefish can't manipulate iridophores and leucophores the same way as it does the pigment-filled chromatophores, Senft said—as far as they know, anyway. So their size, shape, and distribution must be perfectly arranged to generate that whiteness. He shows me images of the insides of leucophores that he's taken, holding both spheres and plates that are a few hundred nanometers across.

"We don't know what the reason is that one [cell] should have one [shape] and one should have the other, because as far as we can tell, they're from the same pool of cells," Senft said. "So we're interested in trying to follow further what the biochemistry might be that leads to one or the other."

Now that we understand the basic components of cuttlefish skin, you might be wondering how they all work together. How do the whites and blues and greens ever get seen if they're covered by the yellow, red, and brown pigmented cells?

These aren't just sacks of moving pigment, as in the chromatophores of other animals, like color-changing frogs. It's a complex, self-contained, sophisticated mechanism. In the cuttlefish's chromatophores, the yellow, red, and brown pigment-filled cells are surrounded by eighteen to thirty muscle fibers (depending on which of the more than a hundred species of cuttlefish you're talking about). These muscles radiate out like the spokes in a wheel, and the cuttlefish can squeeze those muscles, shortening them and pulling the pigmented cell out from a small point to a wide disk, covering a much larger surface area. Now, instead of being a tiny pinprick of red, barely discernable on the surface, it's a broad disk, expanded to 500 percent of its original size. If it "opens" all of the red chromatophores, it will look red. If it relaxes the reds, letting them fall back to pinpoints, and tugs on the yellows in its skin, that stripe will suddenly be replaced by sandy yellow tone. It can do this so fast—inside a few hundred milliseconds—that the patterns seem to zoom across the animal's body. That's why the dark, inky markings I saw running across Buresch's finger-hunting cuttlefish could so quickly erase and redraw themselves.

But then how do we ever see the iridophores and leucophores? Basically, those two layers, full of reflectin-filled cells, are always "on," and they mostly rely on the pigment-filled, muscle-controlled layers above them to block them from view. Some areas, like that disruptive white square in the middle of the cuttlefish's back, have a higher density of leucophores, which is why they can look so blindingly white.

And if you're wondering why the iridophores don't block out the leucophores, well, so do the researchers. Scientists aren't sure to what extent the iridophore plates' movements are controlled—all they know is that these particular cells are not neurally directed the way that the pigment-filled chromatophores are. It seems that the movement of the iridophore platelets is controlled by a neurotransmitter—so that changes happen over longer timescales, seconds or minutes (which is still much faster than chameleon skin).

This question is one of a long, ever-growing list of questions that they have. Take the leucophores: why are the reflectin proteins built into spheres in some places and plates in others? And how does the cuttlefish skin manage to stretch its limited number of pigments out without them fading? After all, it's expanding those chromatophores to as much as five times their resting size, to a point where the pigment granules inside each cell are only three granules deep. And yet, the colors keep their same deep intensity. That's why optical physicists like Hu think that the cuttlefish skin must be fluorescing in some way—generating its own glow. While scientists studying the cuttlefish have shed some light on their ingenious nanoscale mechanisms, there's still much about which they're still in the dark.

Still, the progress that biologists are already making in understanding the cuttlefish's physiology is why engineers like Kit Parker have taken an interest. Imagine this ability, woven into clothing that could change color depending on the environment. In Parker's lab, something of a three-headed chimera where students are working on everything from organs-on-a-chip to grow-your-own-meat to super-fibers, there's a little pink cotton-candy machine that one of his postdocs actually first used to show that nanofibers could be spun with much cheaper processes. One of the students hands me an Iron Man doll whose torso and muscular right calf have been thickly wrapped in what looks a little like that fake cobwebbing that homeowners hang on suburban shrubbery during Halloween. It feels smooth and almost rubbery to the touch. Perhaps one day, if the researchers could figure out how to use color-changing chemicals and proteins, they could potentially weave them into clothing using a spinning apparatus just like that, Parker says.

It's fascinating stuff, outlandish ideas kick-started on a tiny budget (for which the cotton-candy machine stands as evidence), before someone decides the proof of concept is worth putting some money behind. That's the constant struggle, Parker says, though he argues that operating on such shoestring conditions is what forces him and his researchers to innovate. Necessity is the mother of invention.

It might also explain why there are such different biotech-related projects going on in his lab—they're always looking for new opportunities, no matter what direction it takes them. And that's partly why Parker thinks that the first place that cuttlefish-based tech will make inroads is not in combat, but in fashion.

He might be right about that. That's probably where a lot of the money is, he says, walking me across the Harvard campus, on his way to meet a busload of visitors from a company in Thailand interested in the meat-making potential of his synthetic muscle research. (Like I said, his lab is all over the place.) Certainly cosmetics companies like L'Oréal, inspired by iridescent butterfly wings, have harnessed the power of structural color for a shade-shifting line of eyeshadow. "Photonic make-up," they call it.

The cuttlefish's impressive camouflage may have developed as a response to predation, but it has been able to use its skin for a whole other purpose: wooing mates. Males will use their adaptive coloration to create impressive moving bands of light and color that look kind of like a psychedelic sunset in fast-forward, on repeat: pinkish reds and blue hues run across its surface. Even when they're not showing off, they reveal a remarkable mastery of color and pattern (and some canny strategizing, to boot). While studying the mourning cuttlefish, *Sepia plangon*, biologist Culum Brown of Australia's Macquarie University discovered that the males would cross-dress: the half of the male's body that was facing the female would have a large-scale mottled pattern, typical of males; but the other half of its body, facing another male, would be covered in a zebra-striped female pattern! This tactic is thought to give the smaller males a fighting chance to impress the girl before the bigger guys notice. If that isn't a premise for a good animal-kingdom rom-com, I don't know what is.

As I've mentioned before, using their color-changing skills to show off comes in very handy to capture easily awestruck fish and crabs. Go to YouTube and look up something like "cuttlefish" and "hypnotize." You have to see it to believe it. Even as I rewatch it for the umpteenth time, it's hard to fathom how a living being can do this naturally.

The point in swooning over this indescribably beautiful light show is that the animals are harnessing the same mechanism they use to blend into the background to do the opposite: to stand out from the proverbial crowd. It's strangely similar to the way that military camouflage has been repurposed as a fashion statement. Walk down a crowded city street and you're bound to find someone wearing a camo pattern as a fashion accessory: as a jacket, on bandannas, at the beach, on workout capris, even on bikinis. (Even as I sat writing this in a café on Sunset Boulevard in Hollywood on a hot summer day, a girl came walking past my window with two friends, wearing tiny light denim shorts and a green camo-print top that was essentially a glorified sports bra. I don't want to embellish, so to speak, but I have the vague sense that there was also a belly button ring involved.) Camo might be used to blend in, but it can also be used to make a very loud statement. It happens in nature, and it happens in human fashion: the green-and-brown mottled pattern once meant to help soldiers blend in is now used by civilians to stand out.

Fashion seems to have its own set of rules, an internal logic that, at least from season to season, defies physics, economics, and even aesthetics. It does, however, have one interesting similarity to the natural world: strange mutations in an aesthetic appear to pop up every year, and some of them do well and spread, while others die right on the catwalk. In that respect, it's a little—a very little, admittedly—like natural selection. Clothing is like a second skin—and yet we don't take full advantage of it, and make it multifunctional, the way a cuttlefish's skin is. Nature is very good at multipurposing parts, and recruiting one piece of the anatomy for more than one function.

With cephalopods in mind, Parker has taught a class whose undergraduate students teamed up with Los Angeles–based fashion house Rodarte to design interactive dresses laced with fiber-optic light and color. One of the three designs responded to the earth's magnetic field, changing colors as the wearer rotated. Another put on a light show in sync with the speed of its owner's heartbeat; and a third responded to the level of sound in the room. Two of the dresses could even "speak" to each other using radio frequencies, trading colors if both of their owners agreed. (It's an idea that seems to be catching on: A similar dress, made by IBM Watson and designer label Marchesa for the 2016 Met Gala, analyzed the emotions of Twitter conversations surrounding the event and adjusted its colors accordingly.)

Hanlon is also interested in the fashion aspect of cuttlefish camouflage; he's got a high-end, well-known designer visiting his lab the next month, he says, but he won't say who. I wager a wild guess. "Vera Wang?" But Hanlon shakes his head; he's not playing. Fashion is serious business, after all.

More broadly, Hanlon's interested in the art and aesthetic aspect of the cuttlefish. He has co-taught a class held both at Brown University and the Rhode Island School of Design in Providence that brought a mix of students together: artists as well as engineering, neuroscience, and computer science students. One of the major projects Hanlon assigned them: design a two-dimensional background, any kind, and then take a three-dimensional object and paint it so that it blends in. One art student quilted together a tapestry of random triangles and trapezoids and other shapes, all in different shades of blue. Then she made herself the 3D object, designing a dress and painting her face in a chaotic cerulean shattered-glass pattern. Striking as she was, she seemed to blend right into the background. Hanlon, who has had artists-in-residence on and off for the past four years, said she would probably end up doing a stint at MBL.

"How did she get the right shades of blue for her face?" I ask.

"No, no, no, that's not how camouflage works!" Hanlon says. (Clearly I haven't been getting the message.) "You can't just put a uniform color on uniform background, there'll be edges and shadows—that *does not work*. You have to be far more clever than that."

Before we go further, I need to tell you something shocking about the cuttlefish: it's color-blind.

It's true. These masters of disguise, the same animals that can generate a whole palette of shades, from blues to browns to pinks and whites, who seem to blend into the background wherever they wander, whose eyes can actually see polarized light—they cannot see color. Hanlon's group and other labs tested the animals by, for example, putting them in a tank covered in purple and yellow stripes. The purple and yellow were exactly the same brightness, so that there was technically no contrast between them; if you converted the image to black-and-white, the background would look like a stripeless, uniform gray.

The animals did not respond at all to those stripes. They didn't put on a pattern, they didn't raise their arms; nothing happened. Hanlon and colleagues tried again, this time with a large-scale blue-and-yellow checkerboard pattern. Presumably, the animals would put on their disruptive white square, the way they did for the large black-and-white checkerboard. But again, the blue and yellow were the exact same brightness, and instead of putting on a contrast pattern, the animal put on a uniform pattern, as if it were looking at a featureless surface.

The cuttlefish eye, one of the most highly developed light-sensing organs in the invertebrate animal kingdom, is a very different organ from our own. In the retina, humans have photoreceptors called "rods" for black-and-white vision, and cones with three different color-detecting pigments. But behind the cuttlefish's cursive-W-shaped pupil lies just one receptor, which is most receptive to wavelengths of light at 492 nanometers—which translates to a sort of sea-foam green. (That's actually why the researchers used a blue-and-yellow checkerboard for the second experiment: Since blue and yellow overlap with green, it would give the cuttlefish the greatest chance to see the checkerboard, if it could. The fact that it appeared to see nothing, then, seems even more damning now.)

This finding was shocking, for two reasons. One, many of the cuttlefish's predators can see color. Predators of all kinds are highly visual: smell and sound may lead you in the right direction, and even "touch" can be of use—alligators and crocodiles have pressure sensors that pick up subtle disturbances in murky water from prey movements. But when it comes to nabbing a quick-moving meal, eyes are usually what get you the prize.

Two, Hanlon and other research divers have seen, over and over again, how cephalopods—particularly octopuses and cuttlefish—appear to match the disparate shades of the sand, rocks, coral, and algae around them. If they really can't see these colors, it almost seems like a cruel joke by Nature herself—that the animal that relies so completely on camouflage can't predict what its predators (or its prey) may see.

But Hanlon is not entirely convinced that the cuttlefish can't see any color. He and his colleagues Lydia Mäthger and Steven Roberts found something strange: The genetic instructions for opsin, the light-detecting protein in the retina of the eye, were found on the skin of its belly and the tutu-like fin frilling its body. The light-detecting proteins were concentrated around the chromatophores. Hanlon thinks the opsins might have something to do with the animal's incredible matching ability, in spite of being colorblind, though it's unclear exactly how it would work: the opsin proteins are also tuned to a light wavelength of 492 nanometers, which means it would probably not be able to see much other than sea-foam green. But Hanlon's still convinced that there's more going on than meets the eye (so to speak). The opsins appear to be concentrated in the cell types that make up chromatophore organs, along with two or three other molecules found in the eye.

"The equipment seems to be there to detect light," Hanlon says.

The problem is, they haven't yet been able to prove that this is what's happening. Establishing a link between these eye proteins and particular behavior in the cuttlefish has proved highly elusive.

Proven or not, it was an idea that caught the attention of John Rogers, who met Hanlon through a basic research challenge being run by the U.S. Navy and began to read his papers in depth—to a level that surprised and impressed the biologist.

"I mean a lot of people say, 'I see it, I saw your talk,' but they don't read the papers. So when he read the papers, he got it," Hanlon said. "Within five minutes, I knew this guy had it really figured out."

Rogers is a materials scientist and engineer at the University of Illinois at Urbana-Champaign. I'd spoken to him before about the devices he builds, including electronic "tattoos" that, applied to the skin, can monitor your heart, brain, and muscle activity. The man knows how to make circuits that work *on* skin; so perhaps it wasn't as much of a stretch (so to speak) to create devices that work a little *like* skin, too. He doesn't just create flexible devices—he's a flexible thinker, too, one with a deep interest in biology. The cuttlefish was no exception.

"The kinds of things that they can do are just absolutely mind-boggling so as an engineer, it's utterly humbling," Rogers said. "So you look at that, you talk to Roger, you realize immediately, we are not going to be able to replicate that type of system at any kind of detailed level."

Pretty modest words for a guy who was awarded a MacArthur "genius grant" in 2009. Somebody out there clearly thinks that he's got good ideas for the future of electronics. But Rogers says he's just trying to be practical—and to consider what his expertise can bring to the table.

"We can think of it in an abstract sense," Rogers said. "What is the overall architecture, what are the functional layers and how are they talking to one another, and try to pull out those ideas and embed them in the kind of systems that we do know how to make."

Here's the catch with cuttlefish skin: It requires a really big brain. To be able to control a skin that's shot through with so many nerve endings, the cuttlefish must have tons and tons of neurons, which is probably why the Coleoidea—the order of cephalopods made up by the cuttlefish, octopus, and squid—are the smartest invertebrates around. Scientists have put cuttlefish in mazes and watched them quickly learn how to navigate these spatial puzzles.

And they can certainly problem-solve in the real world. If two burly males, competing for the attentions of a female, are engaged in a multicolored camo-off, the smaller males will sometimes put a female pattern on their bodies to get close to the female and mate with her. (And the female often goes for it, apparently preferring brains over brawn.) The clever and cunning octopus tends to get all the academic accolades, but the cuttlefish is no slacker.

If you're trying to build synthetic, skinlike camouflage, however, you don't want to have to carry around a bulky electronic brain (i.e., a computer) to have this thing work. After all, it's hard enough creating synthetic skin that can use adaptive camouflage—imagine having to pair it with a computer processor and cameras. Imagine having to also design and build the programming, and to connect the "brain" and "eyes" to the skin. Imagine all the ways in which that bulky system could break down. This is a fundamental roadblock in actually building a working synthetic model.

Rogers, after reading Hanlon's papers on the potential of opsins, realized something: The opsins provided a solution to the light-sensing problem. If you distributed the light sensing, the way Hanlon and his colleagues believed (but could not prove) was happening, then you didn't need to depend solely on a central hub to watch the environment and react accordingly. The skin itself could do some of that work.

"It really is a bit of genius in my view," Hanlon said. "He looked at the layering and based it more or less as much as he could on cephalopod skin, but he took this next element, which we've been trying to prove for four years but don't have definitive evidence for . . . and he said, 'Now that's something I can do, whether biology does it or not.'"

The prototype Rogers and colleagues came up with was a layered structure, like cephalopod skin. The upper layer's pixels are filled with a temperature-sensitive ink that is black at cooler temperatures but clear when heated above 116 degrees Fahrenheit. This was analogous to the pigment-filled chromatophores. Beneath that was a layer of silvery-white tiles, which could only be seen when the black dye was turned off, analogous to the leucophores. Beneath this layer were thin silicon circuits that could heat up the dye layer to turn it transparent; these were analogous to the muscles that turned the chromatophores on or off. Underneath that was a layer of photoreceptors—akin to the opsin proteins distributed through the animal's body. The skin was able to respond in one to two seconds to changing visual stimuli, from triangles and moving squares to a digital "U of I" logo. The whole thing is a mere 200 microns or so thick, about twice the width of a human hair, and it works even when you wrap that skin on curving surfaces. For now, the device is only in black-and-white, though Rogers says there's no reason they can't ultimately find ways to make their synthetic skin responsive to color.

The list of applications these sort of devices could be good for is enough to make any tech-savvy entrepreneur drool. Such thin sheets could be applied to military vehicles to help them blend in. On the walls of homes and offices, it could be a combination camera and color-tunable wallpaper, perhaps responsive to the ambient light in the room. Rogers talked to his colleagues at the school of architecture, who pointed out that this could be useful for all kinds of surfaces, and not just flat ones—ceilings, tables, other pieces of furniture. The technology may even draw from and perhaps influence the way e-readers and television screens are designed. Like Hanlon and Parker, Rogers has also been talking to a professor of fashion design at the Art Institute of Chicago about the possibilities in fashion and design.

"The disconnect there is that we cannot make a dress. We can make a one-square-inch swatch," Rogers said, with a rueful laugh.

His device is about the thickness of paper, and you can't make a dress out of something the consistency of paper. "She gets excited about stuff and wants us to build pants or something, you know. We'd love to do that—but I'd have to kill a few postdocs before that would happen."

Military applications are just as far off into the future, and not his purview in this research, Rogers added. He, and others also working with the U.S. Navy program, are simply looking at the basic science; this device he created is just a proof of concept that such layered, responsive, color-changing devices are possible. His device only works in black-and-white, but colored component layers could be added in or switched out over time. Using heating circuits isn't the best way to force a color-transition, he added—it wastes quite a bit of energy. But the device gives other researchers a modular template they can use and modify over time, a framework in which to proceed with their own basic research.

All this begs the question, though—does Rogers's innovation count as entirely bioinspired? After all, the opsins' function hasn't been proven; what if that's not how it works? There are long-simmering arguments over what biomimicry means, what bioinspiration means, and what it says about the type and level of science being done, which we'll start to touch upon in later chapters.

Hanlon, for his part, says it does. After all, the insight that led to using the distributed photodetectors only came from studying nature, the biologist says. Whether or not it turns out to be 100 percent accurate is almost beside the point.

2

SOFT YET STRONG

How the Sea Cucumber and Squid Inspire Surgical Implants

I'm hanging out on the sidewalk in a strip mall on a Thursday night. There's a thirty-minute wait for this Korean seafood joint in the middle of Koreatown, the bustling, densely populated neighborhood in the middle of Los Angeles. The lit-up signs are written in boxy Hangul, and the smell of cigarettes fills the air.

The restaurant Hwal Uh Kwang Jang, whose name means something like "fresh square," evokes a farm/ocean-to-table ethos, my Korean American surf buddy told me earlier. That fresh factor is what my friends and I have come for: live octopus, still suctioning your cheek as you swallow it down; lobster whose whiskers still twitch and whose claws are kept securely banded while you dig in.

My friends Swati, John, and I have come here on a mission to try out the sea cucumber, which they serve raw. This is something of a daring turn for me. I'm a fan of Japanese-style sushi, but salmon and yellowtail drenched in mayo and rolling by on conveyor belts are more my speed.

When the sea cucumber first arrives on a bed of crushed ice in a wooden box, I'm grateful they'd only given us a half-order. It's a dark, purplish brown-black color, glistening and gelatinous, a stark contrast next to the bright orange sea squirt pieces laid out in organized fashion on strips of its outer shell. Its pieces lack the clean, calming edges of your standard fish sashimi—an orderliness that I always feel takes the fear factor out of eating something raw and unknown—and instead looks like a series of roughly circular bumps, possibly conjoined, lying across the ice.

Here goes. I take my chopsticks and try to grasp one of the slippery pieces, and bring it to my mouth. It's soft and sort of gooey on the outside, and very, very briny—as if it had already been soaked in salty broth. I finally bite, and I'm almost shocked: I have to crunch through. The sea cucumber is hard—it almost has a fresh snap, like a vegetable. The combination of flavors and textures is weirdly pleasing.

I wonder if I've ever eaten anything that feels like this before, and I think: cartilage, from chicken bones, though cooked chicken gristle tastes soft and crumbly by comparison.

John asks which part of the animal it is. The waitress smiles politely—like many strip-mall joints in Koreatown, there's usually no need for English. He makes a slicing-knife motion with his hands a few times, and then she nods, using her own hands to elegantly describe the process—a slice straight down the length of the animal, open it like a book, whip out the innards, and then take what's left on the insides.

With that last gesture, it's confirmed—we are eating what I came here to try: the inner layer of sea cucumber skin, a material that's as prized by scientists for its strange properties as it is by Koreans as a delicacy.

In the ocean, everybody wants a piece of you. The salty waters are filled with delectable, soft-bodied creatures, from clams to salmon. To go after their meals, fish and other animals with hard skeletons are literally armed to the teeth, which are securely ensconced in bony jaws. To protect themselves against such attacks, other animals like mussels and sea snails armor up with shells, creating a (nearly) impenetrable fortress.

But not every marine resident has the benefit of a shell or teeth. Some, like the sea cucumber and the squid, appear to have missed the evolutionary arms race. Their soft bodies seem woefully underequipped to fend off attack—and yet they manage to survive, and thrive, throughout the world's oceans. That's because these animals, whether predator or prey, have found ingenious ways to incorporate hard elements into their deceptively soft bodies.

That mastery of the hard-to-soft transition is a feat that has eluded researchers working on soft robotics and performing surgical procedures. Scientists (and doctors) need the flexibility that comes with pliable materials, but they also need the structure that comes with rigid ones. Such "smart" materials could transform the practice of surgery, developing better prosthetics for amputees and safer implants for the brain that don't need to be replaced over time.

This chapter will focus on how certain animals navigate this hard-to-soft transition in a complex liquid environment. It's a feat of nanotechnology, one they've each been performing for millions of years, and one that could help doctors and biomedical engineers surmount a major problem: how to build new devices that can be implanted into the wet, soft, salty world of the human body without causing long-term damage.

The sea cucumber is not exactly the beauty queen of the seas. Named for the long, tubular vegetable with a clean smell and a crisp flavor, the animal looks more like a pickle with an unmentionable skin disease. It crawls along the bottom of sandy floors; depending on the species, it can look like a ridged, knobbly, or even spiked giant sea slug. The animal has no brain—just a ring of nerves around its oral cavity that extend both to the tentacles around its mouth and also down the length of its body. It gobbles up the detritus that falls to the seafloor, extending long tendrils that are modified versions of the tube feet it shares with starfish. Depending on the species, these feeding tube feet can be gorgeous and ornate, like the branches in a tree or the delicate ends of a neuron.

These animals hold their feeding arms up to catch particles in the water, or they plow them into the sand and chow down, pooping out the clean sand that covers the seafloor. This trashy diet is probably what gives them that strong briny flavor that makes them a sought-after delicacy in places like China and Korea.

Sea cucumbers are bottom-feeders, a term that's used unflatteringly to describe certain types of people: ambulance chasers, paparazzi, payday lenders. But that's an insult to literal bottom-feeders everywhere. It's true that sea cucumbers eat dead and discarded matter, from carcasses to excrement. But that's not a bad thing—in fact, it's a crucial cleaning service for the world's oceans. Sea cucumbers clean all that crap out of the water and off of the substrate, and then poop out nice, "clean" sandy substrate. They're the earthworms of the sea in that way, recycling decomposing matter and aerating the seafloor.

As the demand for sea cucumbers has grown in recent years, many populations are shrinking, which means more undigested nutrients in the ocean. This reduces the water's clarity—which, for the sea creatures who have to swim through this murky liquid, is probably about as healthy as breathing in the smog of Shanghai. All those extra nutrients can trigger algal blooms, which suck up all the oxygen in the water and cause a mass die-off as fish and other sea life suffocate. Without sea cucumbers tilling the seafloor, it hardens, making it impossible for other benthic organisms to survive there.

Sea cucumbers are indeed the janitors of the ocean. But janitors are often called custodians, and sea cucumbers do perform that function, caring for the ocean that they live in.

Sadly, there's no such thing as gratitude in the wild. While the sea cucumber performs a service that benefits its fellow ocean dwellers, many of those inhabitants see the soft, slow animal as easy fast food. The sea cucumber has a few defenses against predators; some species can shoot their respiratory organs out of their anuses and let them wave around, because the sticky tubes are covered with a soaplike chemical that is toxic to other animals. But it can't hurl its lungs out every time it feels threatened—those organs can take weeks to grow back. Some species burrow into the sand to hide from predators—but that's a time-consuming process, and they can't stay buried forever.

Unlike its cousins, the starfish and the sea urchin, the sea cucumber seems woefully underarmored for the fish-eat-fish world it lives in. Starfish have bony plates made of calcium carbonate called ossicles to protect them, which is why they feel so tough. In sea urchins, those plates have fused together, and its bristling array of sharp spines further warn predators to keep out. But in sea cucumbers, those ossicles seem to have shrunk to near uselessness. This is great for the sea cucumber if it wants to, say, squeeze itself into a safe little nook in a rock or cranny in some coral—it can practically liquefy its body as it pushes into the hole. But that particular quality is not so useful for fending off a razor-toothed attack.

Luckily, the sea cucumber has a secret superpower, one that isn't apparent when it's happily snagging detritus out of the water or pooping its way across a reef. When threatened—and when farting its lungs out doesn't work—the sea cucumber can go rigid, turning from the consistency of play dough to hard plastic. If you're wondering how a soft-bodied creature manages this feat with its shrunken, vestigial calcite plates, you wouldn't be alone.

It's a question that dogged a handful of researchers for decades, and that's because the sea cucumber uses a totally different adaptation from those of its well-known echinoderm cousins.

Instead of relying on its now-shrunken calcite plates, the sea cucumber calls to arms a network of tiny collagen fibers—known as fibrils—embedded under its skin. These fibrils link together, creating a scaffolding through the body that acts like protective dermal chain mail. When the animal is in its soft state, any stresses that travel through a random fibril quickly pass into the soft matrix, easily penetrated by pointy objects (like teeth). But when they're connected, the skin has structure and strength. It's kind of like the beams of a building—you shake one beam, you shake them all, because the force is traveling from one connected post to the other. But even though it's all getting shaken, *the structure doesn't fall down*. When these hard elements are connected, the stress can pass through and be safely channeled away.

The same thing is happening in the sea cucumber skin. Joined together, the collagen fibrils in its skin make a sort of structure through which stress can be transferred safely, without bending or breaking.

But how does the animal do this? That's been a topic of fascination to many marine biologists, but appears to have gone largely unnoticed in the materials science world. That is, until a team of landlubber Case Western Reserve researchers published a paper in the prestigious journal *Science*, where they showed they had created a material inspired by this strange sea creature that could change from hard to soft and back again.

To Stuart Rowan, a professor at Case Western, this research project was a happy accident. He'd been a polymer chemist working on different types of materials with industrial applications as well as more blue-sky research. As a boy growing up in a town called Troon, thirty-five miles outside of Glasgow, he'd explored the rock pools near his home, off the west coast of Scotland, filled with shrimp and crabs and fish and spiral snail shells, a rainbow of earthy tones from purply brick-hued sea anemones to greenish-brown algae-coated stones. But there were no sea cucumbers in the rock pools that he saw—and more than two decades later, in Cleveland, Ohio, the skin of a strange sea creature wasn't even remotely on his research radar.

But Rowan was actually ideally suited for the task. Over the course of his career he'd developed materials with dynamic properties—including polymers that, exposed to a stimulus like light, could repair themselves. These sorts of materials are of growing interest in a field that has often seen materials as static and unchanging once they're produced. Rowan had managed this because he's an expert in what are called "non-covalent interactions," a type of chemistry that deals in the bonds between atoms that aren't as strong as covalent bonds.

For a quick refresher: Covalent bonds hold molecules together by sharing the electrons; the atoms being shared usually are pretty neutral, or close in charge to one another. Ionic bonds, however, are held together by some degree of charge—one atom has an excess of electrons and is negatively charged, another is missing some electrons and is positively charged, and as they do in magnets, opposite charges attract. It's the kind that holds the positively charged sodium and the negatively charged chlorine in your table salt together. These charged atoms, or ions, aren't really sharing electrons the same way as covalently bonded atoms do—they're just fitting together like pieces in a puzzle. Covalent bonds are marriage bonds—shared bank accounts, shared resources. Ionic bonds are roommates—living together but in separate rooms and (mostly) separate possessions.

(That's not to say, however, that covalent bonds are always stronger than ionic bonds—which is what I was taught in high school. Turns out certain ionic bonds can be much stronger than some covalent bonds—it just has to do with the atoms in question.)

There is also a wide range of non-covalent interactions, which Rowan studies and most of which go far beyond the purview of this book. For the purposes of this chapter, we'll be talking about hydrogen bonds, which occur when the electrons in a covalently bonded molecule are *not* shared entirely equally. Instead, the electrons tend to prefer hanging around one part of the molecule over the other. This means that one atom (in this case, the hydrogen) is slightly more positive, and another atom (which this hydrogen is attached to, in the same molecule) ends up slightly more negative—and those atoms can create weak bonds with other molecules that also exhibit this kind of polarity.

Non-covalent bonds—particularly hydrogen bonds—are the cornerstone of fundamental biological processes, Rowan told me. They play a key role in the formation of DNA's double helix; as they break and reform, they allow DNA strands to separate in order to replicate, and then recombine. They help to facilitate the folding of complex proteins, which are the workhorses of cellular life, both the bricks and the bricklayers in the molecular structures of our bodies.

"This is what helps DNA form its double helix, it's what in part helps to control the folding of proteins," Rowan said. "It tends to be reversible or dynamic—it can bring things in contact with each other but they can easily break apart. They're weak bonds, but they're strong enough. People like to sometimes refer to them as Velcro—they come together, but it's easy to come apart."

You may not like to think of bits of your body being constantly velcroed and unvelcroed together, but it's happening, right now. Good news for you: it works exceedingly well.

A hydrogen bond allows the body to be dynamic—to reproduce, to respond to change, to issue instructions. Man-made materials, on the other hand, are static, set in stone. Steel stays hard, rubber stays soft, and the only change you'll see in those materials is degradation over time. But when hydrogen bonds are used to create materials, they allow for a certain amount of dynamism that wouldn't be possible otherwise. Rowan had used his expertise in these non-covalent bonds to create a variety of different materials—"self-healing" coatings whose bonds break apart, allowing material to flow into cracks to repair them; substances that can change color when they sense a dangerous chemical warfare agent; shape-memory polymers that can be twisted into some strange form but spring back to their original shape when heated. They're nowhere near as impressive as a living material like, say, skin, but they have a few aspects of those fundamental properties—responsive to stimuli, and a semblance of self-repair.

Rowan's expertise in non-covalent bonds is the reason that his colleague Christoph Weder, then a professor in the same department, asked Rowan if he was interested in an unusual project. Weder had heard of the sea cucumber's strange morphing ability from Art Heuer, a colleague in a different materials department on campus. Heuer had, for a relatively short time, studied the properties of sea cucumber skin and had wondered if Weder, with his polymeric skills, could actually make such a material.

Rowan was fascinated. He went back to his lab and toyed with different substances, but he and Weder were stuck, because they needed a reason to make these materials. Could it be used for body armor? Perhaps for a Batman-like cape that could turn into a rigid wing? The answer eluded them for months. It was not an idle question: without a clear purpose in mind, without an end product that could be used for military or medical or other commercial reasons, it's hard to get people to fund your project.

The answer came from an unexpected source, in a highly unlikely place: a random meeting whose true nature appears to be lost to the mists of memory—"probably some kind of School of Engineering–level research committee," Rowan said. Some colleagues were late, and to kill time, Rowan started chatting with a scientist from Cleveland's Veterans Affairs hospital, Dustin Tyler.

"We were waiting on people to turn up and we were just shooting the breeze and I told him we had this cool material but we didn't know what we were going to do with it," Rowan said.

Tyler's answer was immediate: "That would be great for cortical implants." Tyler was a biomedical engineer at the VA hospital and had spent his career devising and studying different ways to interface with the brain, particularly in the context of amputees. To him, the usefulness of such a material was patently obvious, and a partnership was born.

"It's the only productive committee meeting I've had in my life," Rowan joked. When it comes to doing groundbreaking research, he added, "half of it is luck."

Another happy coincidence soon came their way, in the form of Jeffrey Capadona. He'd just finished his graduate degree at Georgia Tech, working on chemical surface modifications to biomaterials that would allow the body to more readily accept surgical implants, and was following his then-wife to Cleveland as she started a job at NASA's Glenn Research Center. She passed around his resume, but he was still trying to figure out what to do next—go into industry? Do something in patent law?—when Christoph Weder called him up.

"He said, 'I'm not sure how I got your resume, or why I have it—but it looks like your background is really good for a project we want to do,' " Capadona said.

Capadona had no idea what a sea cucumber was, or why someone would want to use it. But he did just happen to have a budding hobby—raising fish in a saltwater tank—and over time, as his interest in the sea cucumber research grew, he introduced as many as ten sea cucumbers into his aquarium. Currently, he has three or four of them (a yellow one, a yellow one with white spots, and two dark brownish ones) and they do good work, keeping his aquarium filters from getting clogged up with the debris the tank's other inhabitants produce.

The young scientist began pulling up the research literature on these strange animals. The mechanism they needed to imitate seemed pretty straightforward: Collagen fibers suspended in a soft matrix would, on some mysterious command, lock arms with each other, creating a rigid scaffold throughout the skin that gave it sudden stiffness. Then, on command again, these fibrils would release their grip, allowing the skin to relax.

The researchers figured they could copy this trick with a few adaptations. First, to mimic the softness of the pliable sea cucumber skin, they'd use a soft polymer. To imitate the collagen fibers that gave the sea cucumber skin its hardness in the skin's tense state, the scientists used nanocrystalline cellulose—a tough molecule whose strength-to-weight ratio is eight times higher than steel. (Nanocrystalline cellulose is something of a wonder material: supposedly stiffer than Kevlar, potentially electrically conductive, and biodegradable. It could make its way into body armor, glass windows, or even replace the plastics in car bodies. Private companies as well as the U.S. government are trying to produce nanocellulose cheaply without having to kill trees to get it; the United States has built a nanocellulose facility in Wisconsin—which at the time of its opening in 2012 was only the second such facility in the world.)

This hardness, however, wasn't the only reason they turned to cellulose instead of using collagen. The chemistry that activates the sea cucumber's collagen fibers is incredibly complicated, and at the time, was not well understood. But using cellulose in its nanocrystalline form allowed the team to take advantage of hydrogen bonding, the phenomenon that Rowan knew so well.

Cellulose is a carbohydrate, essentially a complex string of glucose molecules that armors plant cell walls, gives apples their crispness and helps make tree trunks so strong. Nanocrystalline cellulose is made up of long, structured chains of cellulose. Like any complex sugar molecule, cellulose is made up of carbon, hydrogen, and oxygen—and several of those hydrogens and oxygens are paired up and dangling off of the molecule. These hydrogen-oxygen pairs make up subunits called hydroxyl groups, and they're key to what makes cellulose stick.

Here's a quick little reminder on why these hydroxyl groups matter. Remember, molecules like water—made up of two hydrogen atoms and an oxygen atom—are inherently polar; oxygen (because it contains more protons, and because the atom is "small" enough that its electrons stay near the protons) pulls on the molecule's shared electrons a little harder, making it a little more negative, and that leaves the two hydrogen atoms a little more positive. This is what's responsible for water's surface tension: a hydrogen from one water molecule will stick to the oxygen of another molecule, which is why water can mound on top of a penny, piling higher and higher, before finally giving in to gravity and dripping off. It's also why hydroxyl groups are found all over sugar molecules, because plants make sugars by breaking up water and carbon dioxide and then using the components to build glucose molecules.

So within a water molecule, each atom is looking for its opposite: the oxygen is seeking out another molecule's hydrogen and the hydrogens are seeking out other molecules' oxygens. Because each oxygen can grab onto two hydrogens, and each of the two hydrogen can latch onto an oxygen, it means there's a potential for four connections in each water molecule (though they're not always all filled simultaneously, Rowan added). If the water molecules were individuals, it would be as if every person in a crowd stretched out both arms and grabbed one leg from two different people. (This of course means that every person will feel two people gripping their legs even as they do the same to two others.) Even though the grip of each handhold is pretty weak, together, this vast, interconnected network of arms and legs will find itself surprisingly difficult to disentangle.

"An individual single bond is not the strongest bond but it kind of works like a zipper, because you have them all the way down the chain. You have thousands, probably millions of them, across all these chains," Capadona said. "They're short-lived, but since there's so many of them, if one pops off it can go right back on, and it's really working as a team that [makes it] much stronger than a regular covalent bond."

These connections are disordered, but it doesn't matter: they successfully create that tough scaffolding we talked about earlier. (They're disordered in the sea cucumber's system as well, which in fact might be an advantage, Rowan pointed out: a random network is better able to handle force coming from any direction.) But how do you turn those connections off, and make the material go soft again? Simple: water. Remember, water is made up of two hydrogens and an oxygen—and it's inherently polar, meaning that the oxygen side is more negative and the hydrogens are more positive. So when the water flows into this polymer matrix, it can interrupt the hydrogen bonds formed between the embedded nanocellulose fibrils and insert its own. With this network effectively broken up by the liquid, the material can then go almost as soft as the polymer it's housed in. (It can't go quite as soft as the polymer because the embedded nanocellulose, even when disconnected, imparts a little extra stiffness.)

This was the idea, anyway. In practice, it took a long time and a lot of thinking to figure out how to make the nanocellulose fibrils—which they harvested from tunicates, another kind of sea creature that uses the fibers to build its expandable protective sheath—disperse evenly in the polymer matrix without clumping together. The big breakthrough came when they realized that they couldn't use two different solvents (one for the polymer and one for the nanocellulose) as was typically done in this kind of work. The trick was to find a single solvent that both properly dissolved the polymer and allowed the fibers to disperse evenly within it.

The result? A polymer with incredible stiffness that could, on aqueous command, become about fifty times softer as the water seeped into the polymer and, more importantly, caused the hydrogen-bonded cellulose fibers to let go of one another.

But the scientists now had another problem: while they needed the material to be soft once it was in position, they also needed it to start out stiff enough so that a surgeon could hold an implant without deforming it and place it properly inside the brain. Simply adding more nanocellulose would make it harder in its "stiff" state, but this would also mean the material would be relatively harder in its "soft" state, too. The scientists certainly couldn't make the nanocellulose fibers any softer, so they turned their attention to the polymer, focusing on its glass transition temperature. The glass transition temperature is a property of amorphous materials, or the amorphous parts within a material. It's when the polymer, when exposed to a certain amount of heat, suddenly goes from stiff to rubbery. This is not a melting point; it's still solid, but any non-covalent bonds have broken up, allowing the polymer chains more freedom of movement and leaving the material very soft. (You can experience this for yourself: Take a piece of chewing gum, chew it up in your mouth until it's nice and soft, and then put an ice cube next to it in your mouth. You'll feel the gum go rigid, even though it's still filled with your saliva, simply because the ice has brought the temperature down below the gum polymers' glass transition temperature.)

Just as different metals have specific melting temperatures, each polymer has its own glass transition temperature. So the scientists switched from an ethylene oxide-based polymer to a polyvinyl acetate whose glass transition temperature was slightly above humans' natural body temperature of 98.6 degrees Fahrenheit. The scientists were taking advantage of both heat and wet working together in the body to alter the polymer's state: Water can reduce a polymer's glass transition temperature just a tad because it can diffuse into the polymer chains, making it easier for those chains to move. So the end result is, when that cellulose-filled polymer is placed in the body, filled with water, and heated to body temperature, it softens dramatically and swells, too—pushing the cellulose nanofibers far apart. The result? A polymer that went from the hardness of a CD case to that of a rubber band, Capadona said—on the order of a thousand times softer.

Since then, the researchers have been trying their jellifying implants out in rats, watching to see how their brains react, compared to standard implantable electrodes. While the response was fairly similar in the first few days after implantation—probably as the brain reacts to the trauma of surgery with inflammation—the brain's alarm systems after two weeks, eight weeks, and sixteen weeks were dramatically lower with the sea-cucumber-inspired implants than they were for the standard glass devices.

Tyler and Capadona kept working on ways to enhance the brain's acceptance of this foreign object, while Rowan and Weder kept thinking of new ways to use this strange material. And Rowan himself now had the bug, the bioinspired bug—he was looking for other sea creatures that could help inspire a new class of morphing polymers.

The advantage of creating a bioinspired material is that now, people start coming to you with the ideas. That's exactly what happened two years later, when Capadona's office mate at the Veterans Affairs hospital pointed to a paper on the squid beak's remarkable set of properties.

The squid, like its cousin the cuttlefish, is little more than a sack of vulnerable flesh with tentacles, making it a tasty target for other denizens of the deep. And yet, these animals are voracious eaters and highly capable hunters in their own right, able to attack and eat all kinds of prey within their size range. Depending on the squid species, their meals of choice can be very small (shrimp and other tiny crustaceans) or very large (enormous deep sea fish like hoki, which can stretch several feet long). And for many species, including the enormous, pickup-truck-sized giant squid, their diet can also include their smaller brethren.

How does the squid devour such enormous, and often fearsome, foes? Hidden underneath its squishy tentacled face is a cruel surprise for the unwary fish or crab—a sharp, hooked beak, shaped like a parrot's. The tip is harder than human teeth even though (unlike teeth) it's all-organic: it doesn't use minerals like calcium to bolster its strength. Dried out, the beak is mostly protein, some pigment, and chitin (the stuff found in insect exoskeletons).

Very few animals can get the best of the giant squid, which can stretch fifty to sixty feet long, but the sperm whale is one of them. However, if you look inside a sperm whale's stomach, you'll find the beaks of long-digested calamari entrees, still intact. These beaks are tough cookies.

As arguably the hardest all-organic material, a squid's beak can slice through flesh and crack open tough shells. Given the squid beak's jellylike body, it shouldn't be able to house a beak like that at all. It would be like trying to hold a blade (sans hilt) in your bare hand. The knife may slice through your intended target, but not before ripping into your own palm. That's why knife blades are secured in safe-to-hold hilts, and why teeth are nestled in strong jaws. That's the only way they can be deployed without doing damage to their pliant, fleshy housing.

Paul Marasco, the former office mate who flagged the research to Capadona, works on medical implants, particularly prosthetic limbs for amputees. A prosthetic fitted onto the stump of an amputated leg allows its wearer to walk, but not without consequences. The hard parts of the prosthetic aren't connected to the bone, but anchored by a series of straps cupping it against the skin. The constant rubbing can prove taxing on the skin and perhaps the remaining muscle underneath. It can be highly uncomfortable to the wearer, and can lead to sores, infections, and further tissue degeneration.

Imagine if you could make all kinds of prosthetic and medical devices with the necessary hard parts—for example, stents inserted into blood vessels—but make their edges soft enough that they play well with the surrounding tissue. Or perhaps you could attach the implant directly to the bone; it could be hard at the bone interface, and soft where it interacted with skin. (The last time he checked, these "percutaneous" implants were not allowed in the United States, Capadona said.)

"The skin just never heals and the infections seep in, and that's why it's illegal in the United States to have that type of implant," he added.

The paper Marasco had shown Capadona, anchored by well-known UC Santa Barbara biologist Herb Waite, had also been published in *Science* in the same month that the Cleveland team had published their 2008 sea cucumber findings. They just hadn't seen it until now.

The paper revealed that squid manage to wield their sharp beaks in their soft bodies by using a network of crosslinked proteins to create a mechanical gradient within the beak. That is, as the beak recedes toward the squid's flesh, it becomes progressively softer and softer, until it's roughly one hundred times softer at the base than the tip. There's no point where really hard material is coming into contact with really soft material, resulting in the kind of mechanical mismatch between glass electrodes and soft brain tissue that the scientists were trying to reduce with their sea-cucumber inspired electrodes.

To find out how the squid manages this trick, the UC Santa Barbara scientists took Humboldt squid beaks (gathered as a spate of the animals were washing up dead on California shores) and cut them into thin cross sections, examining how the gradient changed, slice by slice. They found that the squid beak uses proteins embedded in a soft matrix of chitin, and these proteins crosslink with each other at higher and higher rates. This isn't reversible crosslinking, like the sea cucumber—it's built into the design, the same way the hardness of your nail or the softness of your eye is predetermined. When it's dry, the beak doesn't have much of a gradient, but when it's wet (as it is in the squid's natural environment), that gradient is dramatically enhanced.

Capadona realized that the squid, while using different materials and a different mechanism to manipulate hard and soft materials, was still using the same principle of crosslinking for added strength. It's as if you asked a sea cucumber (assuming you could talk to a sea cucumber) to crosslink zero of its collagen fibers at its tail end, then 10 percent a little farther up, gradually raising the rate of the crosslinking until it reached 100 percent at its front end. And then you basically froze those connections in, so that it would forever remain soft at the tail but hard at the head.

Obviously you can't ask a sea cucumber to do that, for many reasons: It wouldn't understand you and it's not capable. It would be like asking you to tense individual muscle fibers in your bicep just by thinking it. But Capadona knew that if the team could mimic the crosslinking of sea cucumber collagen fibers using cellulose fibrils, then they could also use cellulose to mimic the crosslinking of squid beak proteins.

It was so brilliantly simple. Capadona picked up the phone. "Do you want an idea for our next *Science* paper?" he asked Rowan. "Take me to lunch. . . ."

After lunch, Rowan went back to his lab and told his students they had a new aim in their thesis. By modifying the cellulose fibrils with chemicals that responded to light and introducing another molecule that could link them together, the researchers could use ultraviolet light to crosslink the fibrils and lock in those connections. The longer the light exposure, the stiffer the material was. By exposing each succeeding segment of the material they made to more light, they were able to create a clear gradient in their polymer. The result? A soft plastic that was five times stiffer on one end than the other. That's not quite as impressive at the squid beak's one-hundred-times-stiffer gradient, but it's a start.

These discoveries may seem partly like a series of happy accidents, the product of a smattering of anomalous meetings, but I wasn't so sure. After all, the ideas have a lineage: Rowan mentioned that Weder got the idea from a materials scientist named Art Heuer, who had worked with a biologist who he believed to be named "Jim" Trotter. How long had these ideas been growing and developing until they finally came together?

"I'm actually surprised Stuart even mentioned my name," Heuer tells me when I call him up. Heuer's spent about 49 years on the Case Western faculty and he tells me he's published about 570 papers and had more than 100 graduate students—if only to emphasize that when he says he's surprised, he's not being self-effacing; he's well aware of his place in the academic world. Heuer is technically a professor of ceramics, although he's studied a wide range of materials, from various metals to the properties of eggshells. ("Do you know many devices where, for a dime, you get a sterile container filled with rich nutrients?" he challenges. Chickens, he points out, make them every day.)

Heuer works in a separate department from Rowan—at the Institute for Advanced Materials—while Rowan works in the Department of Macromolecular Science and Engineering. Heuer's department focuses on metals and semiconductors, while Rowan's deals in polymers and what Heuer called "soft-matter stuff."

If you asked Art Heuer what he'd be remembered for among his colleagues, the sea cucumber would not number among them—not by a long, long shot. As far as he's concerned, his contribution is a tiny footnote in Weder and Rowan's head-turning work.

"I had one hundredth of one percent credit for making them aware of this interesting phenomenon," he says.

But around two decades ago, Heuer did study the sea cucumber—and he didn't do it alone. Heuer, along with about twenty colleagues, was part of a remarkably diverse group of engineers, materials scientists, biologists, physicists, and physicians who came together under the Defense Sciences Research Council, part of the Defense Advanced Research Projects Agency. DARPA is basically a research arm of the Pentagon that tries to identify and foster emerging technologies—future tech. DARPA helped birth the Internet: the first message over ARPANET, the network DARPA created, was sent in 1969.

So DARPA knows how to do practical applications. On the flip side, I first covered them when they unveiled their 100-Year Starship program in 2011—an effort to encourage private companies to create technologies that would be useful if (when?) we develop faster-than-light travel that actually allows humanity to reach distant planets. (Faster-than-light travel, for the record, has remained firmly planted in the realm of science fiction.)

DARPA's activities, then, span a wide breadth of probabilities, from ambitious but practical to the kind of half-baked ideas your college pals talked about while fully baked themselves. In this capacious arena, comfortably situated, was the Defense Sciences Research Council (DSRC). This team was essentially the eyes and ears of two other DARPA departments, the Defense Sciences Office (which funds high-risk, high-reward technology) and the Microsystems Technology Office, Heuer said, and so they had to be thinking several steps—and several years—ahead of current technology, and current threats.

"I used to joke the average IQ in the room went down five points when I walked in the room—and I'm pretty well regarded in my field," Heuer said. These twenty or so researchers, he said, were "probably the smartest group of people I'd ever been involved with."

Here's an idea of what the council would do, said Heuer, who was part of the group for more than two decades until 2013, roughly a year and a half before it was dissolved. They'd think up wild-sounding hypotheticals—what if spores of anthrax, the deadly bacteria that can be used as a biological weapon, ended up in the Library of Congress? They'd play out the scenario, work out protocols, write it all up. Then, when an anthrax scare actually did hit a few years later in 2001, "they dusted off our report and there was a whole set of instructions on how to proceed," Heuer said.

Tapping the imagination in service of preparation—that was the DARPA council's way.

During the early to mid-90s, research questions related to biology were not a priority among the various DARPA offices,

Heuer said. But that began to change in the DSRC, and more so when DARPA hired a man named Alan Rudolph (a zoologist by training) who began to look at what insights they could draw in studying nature—whether for biotechnology, biowarfare, and yes, biomimicry. Rudolph ran the Controlled Biological Systems program under the Defense Sciences Office, and collaborated with the DSRC. Rudolph's focus on biology wasn't the only kind of future technology the council tried to identify—but to Heuer, it was certainly the most engaging. Many of the bioinspired technologies that have been making a splash in recent years—gecko-like surfaces that can stick to walls, running robots—were identified by the DSRC, encouraged by Rudolph's Controlled Biological Systems program and supported by DARPA.

"In the mid-nineties they had essentially zero funding for life science and biology biotechnology biowarfare, any of that," Heuer said. "It really took a while before the culture of DARPA changed—but it's changed to the effect of they now have a whole office devoted to this."

That office, the Biological Technologies Office, was just launched in 2014, by the way. "Biology is nature's ultimate innovator, and any agency that hangs its hat on innovation would be foolish not to look to this master of networked complexity for inspiration and solutions," DARPA director Arati Prabhakar told the House of Representative's Subcommittee on Intelligence, Emerging Threats and Capabilities in March 2014.

This change in attitude, Heuer said, was in no small part due to the work of the DSRC, as well as to Rudolph's Controlled Biological Systems program.

"I was the first and maybe the last zoologist DARPA ever hired," Rudolph writes in *Muse,* a blog he writes in his current role as vice president of research at Colorado State University, "and in 1996 I was recruited to use my training in evolutionary adaptation to invest in life sciences to harvest new technological applications (with an eye toward keeping the US technologically superior, DARPA's mission)."

Rudolph's group would put together conferences and invite members of the DSRC, bringing together very different scientists to cross-pollinate ideas across disciplines that, until then, had nothing in common. It was at one of these meetings that Heuer met John Trotter, a researcher at the University of New Mexico medical school. He was, as Heuer likes to put it, a "card-carrying biologist," as if biology were a kind of exclusive club, or perhaps a sort of religious devotion.

Trotter, for his part, was somewhat baffled and a little flattered when he got the invite from Rudolph, who said he'd read his papers on sea cucumber skin and wanted him to present at a meeting that was part of his Controlled Biological Systems program. The title struck him as odd—weren't all biological systems also controlled systems, by their very nature?—but intriguing, and so he hopped on a flight to the East Coast.

Perhaps around fifty people attended the meeting, and their presentations were riveting, Trotter said. One researcher from the University of Montana had used honeybees to map out toxins in an area (which helped nail a case against a company producing pollutants in the Pacific Northwest); a team from UC Berkeley was studying the principles behind the gecko foot to create robots that could walk up walls. Another team from Virginia Tech was dehydrating beetles, sticking them in a vacuum, and then bringing them back from the dead—probing the molecular secrets behind their ability to survive in extreme environments.

"It was fascinating," he said. "I thought this was good company to be keeping."

Trotter was interested in the interfaces between materials at the muscle-tendon junction. How do force-generating filaments within skeletal muscle cells pass the stress on to force-transmitting collagen fibers in connective tissue outside the cells? And within a tendon, how might relatively short collagen fibers and the non-fibrous components work together to transmit stress between muscle cells and bones? Medical colleagues followed his work because they wanted to know when and why such interfaces might fail, which could shed light on tissue damage and healing in humans. Trotter approached these questions by studying sea cucumbers for two reasons: One, because their collagen fibers, unlike those in mammals and other animal groups, can be isolated and tested with relative ease; and two, because the animals could actually change the mechanical properties of their connective tissue using their nervous systems.

The scientist remembered the first time he held a sea cucumber. It was a *Cucumaria frondosa,* one collected while he was on a research trip in Maine. He picked it up and felt it stiffen in his palm—watched the way the skin seemed to flow between his fingers as the animal relaxed, the impression of his fingertips lingering in the flesh.

"I was hooked," Trotter said. He had to find out how the sea cucumber did it. And after years of painstaking research, he had finally figured out some of the basic mechanics of the skin—research that he was presenting at the Controlled Biological Systems meeting.

Art Heuer, who was also attending the meeting, was blown away by the sea cucumber skin's mutable mechanical properties, unlike any man-made material then known.

"I'd never even thought of the word 'mutable,' says Heuer, who works with hard materials. "I knew the word 'immutable.'"

Rudolph had connected Trotter and Heuer at the meeting because, while he encouraged Trotter to write a white paper (a precursor to submitting a grant proposal), he couldn't help him for ethical reasons. And Trotter, a biologist, had never gone through the DARPA process before, which is very different from the process for life sciences researchers by the National Institutes of Health.

Heuer was fascinated by this strange material, as was his fellow council member Milan Mrksich, currently of Northwestern University. He brought sea cucumber dermis to Case Western and subjected it to the rigors of his universal testing machine, a device that can probe the limits of a material's compressive or tensile strength, by pulling it or crushing it until it snaps or breaks. While Heuer examined the skin's mechanical properties, Mrksich focused on the biochemistry.

"It was interesting stuff," Heuer said. "We found things that biologists didn't know—but the language we used to describe what we were doing must have been too unfamiliar."

Heuer tried several times to get a paper published on the sea cucumber skin's mechanical properties but the paper never quite made it into a journal. To be clear, Heuer's examination of the sea cucumber wasn't his first time around the research block—he was an experienced scientist with hundreds of papers to his name. There was a deep-seated divide, and his research fell in the gap between them: the biology journals didn't know what to make of his materials science approach, and the materials science journals didn't see a use for the biology work.

"It was one of these things," Heuer said. "Maybe it might be different now."

Trotter, for his part, had been just about to shut down his research lab when he got the call from Rudolph. If it hadn't been for that call, he never would have ended up with funding for three more years to study the sea cucumber dermis. And without meeting Trotter, Heuer would never have passed the idea on to Weder and Rowan, who would have never thought to make such a strange, new, bioinspired material.

Trotter says that when it comes to the chemical complexities of how exactly the sea cucumber works, scientists still only know part of the picture. At the end of those three years, he closed down his lab and finally turned toward his long-term plan of joining the administration at the University of New Mexico in Albuquerque.

"So I regret that when I think about it," Trotter says with a little laugh. "But on the other hand, I don't know that I would have been able to get funding to continue working on this; I think the DARPA thing was a kind of fluke."

I'm not so sure it was a fluke, though. I think the scientists were stymied by issues of translation—researchers had to learn to speak the other disciplines' language, to get a sense of their values and their culture. To be extremely reductionist: Biologists often want to study a system in painstaking detail. Materials scientists and engineers often want to help build useful products. Both of these end goals are profoundly valuable—and they're both necessary to take deep, insightful lessons from the world of nature. But in order for that to happen, scientists need to understand one another first.

It's sort of like the connectivity in the Cleveland team's bioinspired polymer. After all, there's a certain critical mass of collagen fibers you need to make this hard-to-soft polymer. Too few, and the network will be patchy, with gaps in between. But with a high enough density of fibers, those connections can be fully made. It took a decade before that critical mass of research and researchers grew until it was possible to for them to make the figurative connections and develop this final product.

Rowan humors the metaphor when he hears it. There are certainly issues of translation between different types of scientists, he says—which he still comes across in his work today. "And they all have funny accents," the polymer scientist jokes in his native Scottish brogue.

The day before I flew to Namibia to hang out with scientists studying termites—fascinating work I'll be telling you about in chapter 5—I watched a three-year-old undergo brain surgery. His name was Auguste, he had big blue eyes and a cheeky smile, and he was born deaf. Thus far nothing—not a hearing aid, not even the cochlear implants embedded in each ear—had helped him.

A team of doctors and researchers had chosen Auguste as their first patient in a clinical trial for a device called an auditory brainstem implant. Surgeons would have to cut through skin and bone and the brain's protective sac known as the dura, drain brain fluid, gently push the cerebellum aside, and finally reach the brainstem. There, they would have to find a specific nook in the tissue folds where they would carefully place an electrode array that would be able to directly stimulate the auditory nerve.

It was a six-hour operation—one made longer by an agonizing period when the special spot could not be found, and the doctors who were deep inside Auguste's skull may have privately considered what it would mean to have to abort the surgery on the first patient in their clinical trial. When they finally found the spot, and carefully placed the implant, there was a palpable release in the room. A month later, the researchers tested the device. When he heard his first sound Auguste froze, and looked up.

I watched the surgeons work on this little red window, tried not to think about passing out, and thought, "Surgery is a messy business." The surgeons cutting into the skull really had to put their shoulders into the work. When they opened up that little red window, you could see the brain—soft, pulsating, jellylike. On a visceral level, given the sharp metal tools and the defenseless pink tissue, the fact that the brain was able to take that kind of invasive procedure still amazes me, many months later.

The brain heals from the trauma of surgery. But assuming the procedure has gone well, the brain has to deal with this little metal-and-glass invader, the electrode array, for the rest of its existence. Over the weeks following the surgery, fibrous scar tissue starts to form around the metal, the body's way to keep it from damaging surrounding tissue. That scar tissue, while necessary, reduces the effectiveness of the device that the surgeons just worked so hard to implant.

Auguste is lucky; his electrode array sits in a spot where there's little movement, so the hard electrode doesn't rub too much against the soft tissue. But other chronic neural implants come with a higher price down the line. This is no knock against any of those implants: Some are used to treat the seizures due to epilepsy; others are used to treat the shaking that comes with Parkinson's disease. That said, studies show that the efficacy of these devices can degrade over time, as the brain tries to deal with the constant inflammation caused by this foreign object. They have to be removed and replaced, which means patients have to allow themselves to go under the knife over and over again.

Surgeons, then, are stuck between a soft and a hard place. They have to work with rigid, unforgiving materials that do not mesh well with the body, physically or chemically. Researchers are coming up with a number of different possible solutions, from biocompatible molecules to printing circuits onto polymers as thin as plastic wrap. It's unclear which will prove the best—perhaps a combination. But consider this: the sea cucumber and squid beak operate in salty water, and so must the devices that work in the human body. Some of the most promising micro- and nanoscale solutions for pressing biomedical problems might, at this very moment, be sliced up on dinner plates around the world.

PART II

MECHANICS OF MOVEMENT

3

REINVENTING THE LEG

How Animals Are Inspiring the Next Generation of Space Explorers and Rescue Robots

Few robotic deaths have seemed as tragic as the death of *Spirit* on Mars.

The NASA rover, described in turns as "plucky" and "brave," was launched in 2003 with its twin *Opportunity* to study the Red Planet's surface. The rovers were the successor to the Pathfinder mission, which in 1997 landed and released the toy-sized *Sojourner* rover onto Mars. *Sojourner* was also the first successful rover to roam another planet's surface and the first to nearly "break the Internet," as it were: the *Mars Pathfinder* site racked up roughly 220 million hits within four days of its July 4 landing—impressive by today's standards and absolutely wild in those pre-Y2K days. The overwhelming interest was a testament to the growing fascination with a once-alien planet that was starting to look like a near-Earth twin: one that, through a few circumstantial details, had suffered a much drier, deader fate.

But the rovers were also launched at a time when other missions to Mars had literally crashed and burned—though not necessarily in that order. The year 1999 was a bad one for NASA: the agency had suffered back-to-back failures with the Mars

Polar Lander and the Mars Climate Orbiter. The orbiter's end was as exquisitely embarrassing as it was traumatic; the ill-fated spacecraft burned up as it unintentionally passed through the atmosphere, essentially because someone at Lockheed Martin had programmed the spacecraft using English units of measurement instead of the metric units that are the scientific world's standard. Other nations' efforts were faring no better; Japan's Nozomi Mars orbiter mission was aborted because of equipment problems, and England's *Beagle 2* lander completely vanished during its Christmas Eve 2003 descent.

So it was with great hope and great trepidation that the team at Jet Propulsion Laboratory (JPL) watched *Spirit* and, three weeks later, *Opportunity* make their dramatic descents toward the surface (known then as the "six minutes of terror.") When *Opportunity* sent its first signal back from the planet's surface, the JPL control room burst into cheers, and amid the hugging and high-fiving engineers, one of them grabbed Rob Manning, chief engineer for the entry, descent, and landing team, around the shoulders and shook him so hard that he had to hold onto his glasses during the joyous thrashing.

The successful landing may have seemed nothing short of miraculous, but the following days, and weeks, and months, proved an even greater miracle. *Spirit* and its twin were only slated for a ninety-day mission—and the little rovers far outlasted that lifespan, discovering dramatic evidence that Mars rocks were once filled with running water, traveling far beyond their landing sites, living ten, then twenty times longer than their original ninety-day missions. To date, *Opportunity* is still roaming the planet, digging up signs of water and organic-rich sediments, having lasted more than forty times its planned lifespan.

But every streak of good fortune has to come to an end at some point. For *Spirit*, it was one of the wheels that finally did the little rover in. In May 2009, driving through a region called Home Plate in Gusev Crater, its wheel broke through the thin, crusty surface and plunged into soft sand. It soon became clear that digging itself out was not an option: The more the rover spun its wheels, the more mired it seemed to become. The scientists tried to roust the rover but in vain; they finally called it quits in 2011.

"We have developed a strong emotional attachment to both of these rovers . . . they are just the cutest darn things out in the solar system," John Callas, the rovers' project manager at JPL, said in a press briefing at the time.

When the massive, state-of-the-art *Curiosity* rover landed on the Martian surface in 2012, practically every member of the mission control crew leapt to their feet, cheering and high fiving. That rover is the sexiest thing on six wheels on another planet. It sports a suite of laboratory instruments in its belly, a rock-zapping laser in its eye, and a drill that can bore into and sample rock—all of which it has used to show that there were life-friendly environments on Mars in its distant past. But even this advanced rover, far bigger and badder than *Spirit* and *Opportunity,* was almost done in by its wheels, when an image in late 2013 revealed alarming gashes in its aluminum tires. The rover had to moon-walk across one dangerously jagged surface, driving backward just to even out the damage.

The rovers we send to other planets have always been wheeled. The Mars 2020 rover is set to use almost the exact same template as the *Curiosity* rover, perhaps even its spare parts, even though its mission is supposed to be quite different. There are some ventures, after all, in which you do not want to take any more risks than you absolutely must, and wheels are a tried-and-true technology. Why spend the incredible amount of time, money, and stress to come up with another design, when you already have one that works? It just doesn't make sense.

But it turns out, sometimes the very wheels we've depended on for more than five millennia are themselves a weakness; they get stuck in sand pits or slip on rubble-strewn inclines and are easily damaged by unexpectedly sharp, pointy rocks (particularly for a heavier vehicle like *Curiosity*). Wheels, in short, can put the "liability" into "reliability."

This isn't just an issue on distant planets. Engineers of all stripes are looking to non-wheeled robots to be useful in a variety of places on earth, but perhaps most urgently in disaster zones—earthquake-ravaged areas or chemically contaminated buildings, where it's too dangerous to send humans and ineffective to send wheeled robots to help perform recon or even do simple tasks during search and rescue missions.

Wheels are fine for clear, paved roads. But when the going gets rough, you're gonna need to leg it.

Deep within the grounds of NASA's Jet Propulsion Laboratory, past docile deer munching on fallen jacaranda blooms and a faded street sign that reads ROVER XING, Brett Kennedy and Sisir Karumanchi stand in a narrow parking lot, watching what appears to be a strange white suitcase sitting on the asphalt. Nothing happens; I shift from foot to foot. But then, on some unseen signal, the luggage comes to life.

Four limbs, each with seven joints, extend from its sides, bending and clicking into position. Two spread out like legs and two rise up like arms as the robot goes through several poses, looking to my mind like a Transformer doing yoga. The process is slow, but deliberate, and finally the strange creature turns and faces the structure looming behind it: a door built into a wall of chipboard, leading to a three-sided "room," which looks rather like a theater stage.

This is RoboSimian, a prototype rescue robot that Kennedy

and his fellow JPL engineers hope can win the $2 million top prize at the DARPA Robotics Challenge (DRC) taking place a week from today in Pomona, California. It's a contest that has been three years in the making, with an ambitious goal: to motivate the best and brightest engineers around to world to build the next generation of rescue robots—the kinds of machines that could help save lives when the next natural or man-made disaster strikes.

The competition, which I'm covering for the *Los Angeles Times,* is an obstacle course; the chipboard stage is a best-guess version that the engineers have set up to practice with. It reminds me of the television show *American Ninja Warrior,* except with fewer muscles, fewer moats, more machines, and more minutes. Where the humans on *American Ninja Warrior* have sixty seconds to leap and swing and climb their way through the course, the robots have sixty minutes to drive and walk and roll their way through.

Unlike in the adrenaline-fueled TV show, the robots' gauntlet won't require hopping from log to log over a swampy water or hurtling across a chasm on a rope. Here, the robots must do mind-numbingly normal things, like opening a door.

Before it gets to that door, each robot will have to drive a car—maneuvering it around barriers on a course to get to that chipboard room. You'd think that would be the hard part, given that humans have to wait until they're nearly adults to get a permit to drive a car; it's a pretty complex skill set.

But not so. The hardest part of that drive, says Kennedy, won't be sitting behind the wheel, but getting out of the vehicle. That's because, of the twenty-four teams from around the United States and the world set to compete, the vast majority of the robots will be some kind of anthropomorph, with two arms and two legs. After all, the robots will be required to do human tasks, like navigate rubble and operate power tools, so it makes a certain amount of sense to give them a humanoid form.

Here's the hiccup, though: two-legged balance is not easy—which is probably why we're the only primates to walk primarily on two feet.

It's hard enough for these robots to put one foot in front of the other, but having to step from a car all the way down to the ground without falling? They may be able to stride across the safe, predictable confines of the laboratory, but when exposed to rough terrain or even a little wind, they can quickly stumble.

"The thing that would almost guarantee our winning the program," Kennedy tells me as he watches the rover slowly inch its way through the faux obstacle course, "would be if the Santa Anas kick up."

I can't quite tell if he's serious, but he's put his method where his mouth is. Kennedy's team has purposely veered away from a purely humanoid form. RoboSimian, as the name implies, takes a different tack, drawing inspiration from our fellow primates, which can knuckle around on all fours. Humans' arms and legs are very different pairs of limbs—legs are clearly for walking around, arms are for reaching out and grabbing things. On one of our distant cousins, such as an orangutan, the arms and legs are not so different—long, prehensile, both able to support the body while also good for swinging from trees.

In that way, RoboSimian is more primate-like than the other humanoid robots, because all four limbs are essentially the same; this allows it to move on all fours or switch to its hind quarters and use its top limbs as arms.

"We do have a poop-flinging behavior," Kennedy joked, referring to the way the robot uses its arms to knock debris out of its way. "I don't know why no one else wants to call it that."

Kennedy gets into the vehicle that RoboSimian will drive, a Polaris Ranger that looks like a cross between a smart car and a Jeep, and grips the top of the doorframe to show how the rover will exit the car. The robot's pincerlike hands will grasp the frame while it steps out, so that if it does slip, it'll still be hanging on to something, monkeylike, and won't fall flat on its face. Thus, it's able to avoid dealing with that pesky issue of balance to the degree that the other robots will. (It can also wheel around, choosing a best-of-both-worlds strategy.)

Other teams made modifications to their vehicles—ripped out the interior cushioning to make more room for the oversized robots, or add a little step to make disembarking easier—but Kennedy sniffs at the idea. After all, if you're sending this robot into a disaster zone, and it has to commandeer a nearby car, it won't have time to customize its ride.

RoboSimian is not the first DRC contestant I've visited this day. Earlier in the morning, clear across town, I stopped by the lab of UCLA engineer Dennis Hong. Hong bounces around his lab space like an Energizer Bunny, showing off the menagerie of mechanical critters he's built, mostly while at Virginia Tech, where he worked until a year before. Hong points to a spideresque three-legged robot, a snake robot, and even an amoeba robot. None of these, however, will be stepping onto the obstacle course in the coming days.

Hong introduces his entrant into the DARPA Robotics Challenge—a two-legged competitor named THOR-RD. One of two models sports yarnlike pink hair shooting out of its thin head, reminiscent of Beaker, the Muppets' hapless lab assistant. Unlike RoboSimian, THOR has distinctly human proportions—which I find surprising, given Hong's apparent love of unconventionally shaped robots.

Hong is quick to affirm that he's a big fan of animal-inspired robots. Several years ago, during a stint at Jet Propulsion Laboratory, he even worked with Kennedy on LEMUR, a monkeylike predecessor to RoboSimian. But when it comes to dealing with disaster situations, he thinks the human form is the best one. For one thing, if that disaster involves human structures, you'll probably need something with a human shape and limbs to manage doorknobs, climb stairs, and reach for high-up objects.

On top of that, he adds, all these animal-inspired robots tend to have one purpose, or only be useful in a limited number of places. The snake robot would be good for tunneling through debris, but not for unlocking a door. A spiderlike robot might not be your best bet at driving a car. A human robot is like a Swiss Army knife and could perform all of these tasks—theoretically, anyway.

For that reason, Hong seems a little skeptical of the path that Kennedy and the other RoboSimian engineers have taken.

"If they fail, it means, oh, what I said was right," he told me earlier that day. "And if they indeed do well, it means they've proven me wrong."

A few days later, I meet Gill Pratt, program manager for the DARPA Robotics Challenge, out at the Pomona Fairplex, where workers are busily constructing the four parallel stages in front of the grandstand that will house identical obstacle courses. Twenty-four teams are in the running, and four teams will compete at a time. Giant Jumbotron screens will switch between stages, so that audience members sitting in the bleachers or at home will be able to catch the robots' successes (and their epic fails) in larger-than-life detail.

The idea for this challenge—whose top three performers will win a $2 million, $1 million, and $500,000 prize, respectively—sprang out of the 2011 earthquake and tsunami in Japan that led to a disastrous meltdown at the Fukushima Daiichi Nuclear Power Plant. The explosion, the thinking goes, could perhaps have been avoided if someone had been able to get into the plant to open some valves to release the steam that was building to dangerously high levels. But because of the massive amounts of leaking radiation, employees could not get close enough, fast enough. If there had been able-bodied robots that could have performed such a simple task, perhaps they could have avoided the explosion that damaged reactor unit 2 and led to radioactive water spilling into the ocean for days.

As of yet, that technology does not exist. But DARPA is seeking to jump-start these kinds of emerging technologies. One way they do that? By putting a big prize out there. There's nothing like a little friendly competition for a healthy financial incentive to get people to solve a problem.

The robots will start their engines, maneuver the Polaris vehicles around barriers as they drive, and then ditch the car in front of the stage's entrance: an unassuming front door, with the words CAUTION: HIGH VOLTAGE. DO NOT ENTER THIS ENCLOSURE looming above. Once they open the door (if they manage to do so without falling over) they'll face a zigzag wall of faux exposed brick and corrugated iron, accessorized with more warning signs and lined with different tasks.

Among them: The robots will need to turn a valve, drill through some dry wall, and either climb over a pile of cinder blocks or shove their way through some debris. There's a spot in the wall for a surprise task, which will change for each day in the two-day competition (and which Pratt refuses to tell me ahead of time).

That's why both the JPL and UCLA teams have each been practicing with a variety of possible tasks in their faux obstacle courses: pull a triangle-shaped handle; press a button. (If you don't think a surprise task is a big deal, consider this: Even the smallest changes in movement can mean the difference between success and failure for these robots. For example, during JPL's practice with RoboSimian, while reaching for a drill to cut through a wall, the robot's operators reached for the drill on the lower shelf instead of the higher shelf, as it usually did. They promptly knocked the drill off the shelf.)

Engineers have made figurative and occasionally literal leaps and strides in the last decade when it comes to building increasingly athletic robots. Robots can now run faster than Olympic sprinter Usain Bolt; Honda Motor Co.'s Asimo robot has played a little soccer with President Obama.

But many of these robots only do well in controlled circumstances, within the safe, predictable confines of the lab. This is especially true for bipedal, or two-legged, walking robots, because the highly tuned balance needed for this feat is not so easy to replicate in metal and plastic. The materials are far from perfect, and the software is not quite sophisticated enough to make up for it.

To make matters worse, once the robot steps through the door, Pratt and his team will cut out communications, making it nearly impossible for the robots and their human operators, sitting in a distant garage, to speak to each other. This is exactly what happens in a disaster zone. If you're an operator planning to joystick your robot through, say, a hospital struck by an earthquake, good luck to you, because life is not a video game. Radiation could jam your communication with the rover. The building itself may block incoming signals (as anyone who's tried to use a cell phone in a hospital can attest). And there may be so many other people also trying to get a call or a signal through that it may clog the system. Back in the faux bunker, the robot's operators will be experiencing possibly the worst video game connection of their lives; they'll basically only receive about one second's worth of data for every thirty seconds of their allotted hour.

"If you've ever been on a really bad cell phone call . . . it's like that, but ten times worse," Pratt told me.

The problem with such limited communication is twofold: If the robot can't send information back to the human operator, theoretically hunkered down somewhere far enough away to be safe, then the operator has no information to work with, and no guidance to give the robot.

This isn't some elaborate hazing ritual. Pratt and his colleagues want to see which team's software is best able to operate on its own. There's a certain amount of autonomy that's needed if a robot can't rely on constant, immediate micromanaging. Now, we're not talking full-on artificial intelligence. But if a robot sees a branch on the ground, it knows it can clear that branch while it waits for instructions, or—if the connection cuts out partway through the valve-turning exercise, for example—it knows how to complete the motion (and the task).

This is where the NASA RoboSimian folks think they'll also have a distinct competitive advantage, because they've already operated space robots in this manner for years. Since the Martian day is out of sync with earthly schedules, the engineers in the JPL control room often have to wait several hours, or a whole day, to hear back from the Mars rovers, and so they've developed software that gives the robots enough autonomy to perform certain tasks when they're out of touch.

Many teams performed fairly well during the DRC semifinals in late 2013—but now, DARPA has upped the ante. For one thing, the robots can't be connected to power cords; they need to carry all their bulky batteries. They also can't rely on safety belays to keep from falling—which won't be fun for the two-legged robots, which struggle with balance (and because they're so vertical, risk serious damage if they do fall). This is where the wheeled and four-limbed contestants may have an advantage; they're inherently stable.

"In a real disaster, there are no ropes to hold you up," Pratt says.

Pratt doesn't expect all the robots to make it to the finish line.

There are a total of eight points that can be won, and the robotics teams can attempt them in any order they want. If they don't have the ability to get behind the wheel, they can skip the drive, for example, and just hoof it to the front door—but they lose the opportunity for two points, both the driving and the disembarking. And traversing the rough, sandy turf of the field is no walk in the park.

Pratt's plan isn't just to reward the best and brightest robotics team, and to push the field forward, but also to show the public that these robots have a long, long way to go.

When people think of robots that can do human tasks, they think of the technology envisioned in movies like *Terminator* or *Iron Man*—smart, agile, incredibly complex, very robust, and often, highly verbal. If you watched the DARPA Robotics Challenge, either in person or on the live feed, you know that that is far from reality. These robots literally fell apart when trying to do the most basic of tasks, including getting out of a car and opening a door—tasks you or I (or even a three-year-old) would have no problem with.

"We have what's called 'the C-3PO effect,' which is, 'Why isn't this thing doing what C-3PO can do?' " Kennedy explained, referring to the neurotic android of Star Wars fame. "Because C-3PO is more fantasy than science fiction, that's why."

Dennis Hong is a self-professed Star Wars nerd: He fell in love with the idea of robots at age six, after watching the space epic at the then-Grauman's Chinese Theater on Hollywood Boulevard, during a family trip to the states. Star Wars might be part of the reason he's in robotics at all—and yet, even he agreed that such media depictions set the expectations too high.

"They're expecting robots to run, lift things with one hand, they watched *Iron Man* . . . and these robots walk like this," he said, doing a slow-motion shuffle forward with his arms tucked tightly into his body, making C-3PO look like a champion sprinter.

When the robots fall and break when they're trying to do something as simple as get out of a car, "people are very disappointed to see that—but that's the true state of the art," Hong said. "This is not a bad thing; it helps them recalibrate their expectations."

Pratt in particular has a certain depth of perspective on how far robotics has come, and so can take the long view on how much further it has yet to go. In the 1990s, at the Massachusetts Institute of Technology's Leg Lab, he invented the series elastic actuator—which helped to make cheaper legs on walking robots a reality.

Roboticists had long struggled with making legged robots for several reasons, according to a paper Pratt penned in the journal *Integrative and Comparative Biology* back in 2002.

"For both historical and technological reasons, most robots, including those meant to mimic animals or operate in natural environments, use actuators and control systems that have high (stiff) mechanical impedance," he wrote. "By contrast, most animals exhibit low (soft) impedance. While a robot's stiff joints may be programmed to closely imitate the recorded motion of an animal's soft joints, any unexpected position disturbances will generate reactive forces and torques much higher for the robot than for the animal."

Depending on how you define a robot, our relationship goes back centuries, perhaps even millennia. Leonardo da Vinci drew plans in the 1400s for an armored knight (which NASA roboticist Mark Rosheim successfully built in 2002). But in practice, there was little in the intervening centuries that elevated mechanisms to the vaunted status of robot. They had the hardware, but software would not come until the twentieth century, and the age of silicon.

Even back then, human imagination far outpaced the available technology. When seventeenth-century French philosopher René Descartes redefined the human as mind and body, he compared the body's workings to a machine—though a very complicated one. The implication, in reverse, is that a machine could potentially take on lifelike, even human, qualities. This may have helped set the stage for the "automata" of the eighteenth century, built by cunning artificers for the entertainment and use of royalty, clergy, and nobility. Engineers like Jacques de Vaucanson seemed to embrace Descartes's ideas: In 1737, he built the Flute Player, considered the first fully biomechanical automaton. He used bellows to give it breath and reportedly even used skin on the flutist's fingers to give it a gentle touch (though what kind of skin, or whose skin, remains unclear to me). Later, he built the Digesting Duck, a mechanical quacker that appeared to consume, digest, and defecate its meals.

But the humanoid automata were typically fixed-base: They didn't have to move or interact with their surrounding environment in any way. This meant that the makers of automata could use precise, pre-calculated movements to control their position in space. Industrial machines in the nineteenth and twentieth centuries, particularly the welding and painting robots in factories starting around the 1960s, followed the same pattern.

"For these tasks, holding accurate position in the face of force disturbances was paramount and the force necessary to accomplish the task was irrelevant," Pratt wrote.

By controlling their position and by having incredibly stiff parts, these robots were able to machine metal to an accuracy of less than a thousandth of an inch, he pointed out.

Because of this history with what engineers would call "high-impedance robots," Pratt writes, "most designers of today's walking robots (particularly those in industry) come from a background rich in the history of high-impedance mechanisms and control."

The problem is, position-control mechanisms have profound limitations for robots, because they work as if the environment doesn't change, as if there's no uncertainty in motion. In the real world, that's far from the case. Imagine trying to shake hands with a robot. If that robot squeezes to a predetermined position with every single hand, regardless of the size or shape or the strength of the grip, you can bet it will crush many fingers without knowing the difference. A better way to handle motion, when interacting with an unpredictable environment, is to base your movements on how much force you're exerting on a given object.

This is how biological systems—human legs and arms, for example—actually work. If you try to push open a swinging door, for example, you'll first reach out, make contact with the door, and then once you feel enough pressure to show you've made contact with it, you'll push against it. If you try to push without feeling the door first, you'll likely end up accidentally slapping it (and hurting your hand) instead.

Humans and other animals move lightly though the world—we walk, run, and jump (usually) without injuring our ankles and knees. A legged robot that tried to perfectly calculate the position of every step would be hard-pressed to keep up with us—or would risk miscalculating and blowing out its own joints.

Instead of gears and motors, living vertebrates use soft, pliable muscle to move around. Those muscles are attached to our hard skeletons with tendons, which are tougher than muscle but still flexible. These soft elements essentially allow for a little room for error, allowing the body to store and dissipate the force of impacts. But muscles aren't just motors—they also act as sensors, getting constant real-time feedback from their movement.

Many engineers tried to add soft parts to their robots' appendages superficially, by covering hard ends in soft plastics. But Pratt's series elastic actuator was inspired by natural properties. He wanted to create an effective force-control mechanism, but such motors are often expensive and complex. Pratt solved this by drawing from the idea of a tendon. He put a spring in the motor mechanism, right after the gearbox. This innovation allowed for a certain amount of flexibility, which takes the burden off of the software to calculate that next step perfectly. Better yet, the spring can also act as a force sensor—and thus serving, as muscles do in nature, a dual purpose.

There are many types of robotic "muscles"; Pratt's series elastic actuator was far from the first. But his invention made such a compliant, force-sensitive joint actually affordable, and helped foster a boom in legged robotics.

Still, when it comes to harnessing the right blend of materials to get the right balance of stiffness and flexibility, walking robots' hardware still has a long way to go.

And then there's the software. A human standing on one leg isn't just sensing pressure in order to balance; they're also using their eyes and ears, making subconscious calculations. Try this simple exercise: Stand on one foot on a grassy knoll or something. Feel fine? Now close your eyes. If you're anything like me, you'll quickly start to wobble. That's because, unbeknownst to you, your eyes were checking out your surroundings and your brain was gauging how much you needed to tense your muscles to stay upright.

No matter how compliant a robot's joints are, the software needs to do a certain amount of complex calculation, using feedback from cameras and other sensors. That software is getting better and better, but, as the DARPA competition revealed, there's plenty of work left to do.

About a five-minute walk from the grandstand, a cavernous garage is a bustle of activity, as forklifts rumble by bearing enormous crates and graduate students push three-seater couches into place and set up giant computer screens. Each of the twenty-four teams gets their own berth, and it's from here that they will operate the robot, without being able to see the course in real life. This is how it would feel for actual rescue workers using robots in real-life situations—everything will be interpreted through a computer screen.

Robots are unpacked, set up, and calibrated. The team from Carnegie Mellon University hoists up CHIMP, their red, 443-pound robot, to give it a thorough checkup, and they warn me to take a step back. If one of those legs suddenly twitches, it would be like getting sucker punched with an anvil. (That is, for the record, one of the arguments for building more biologically inspired robots: they would be softer, flexible, and less likely to hurt the humans they're supposed to help.)

"You can see all the dust all over the robot from practicing so much—lots of scratches, lots of scrapes, getting run hard," said Eric Meyhofer, the team's technical lead. "CHIMP's been running one hundred hours a week."

Like RoboSimian, CHIMP has apelike characteristics, including long arms it can use either to clamber over obstacles or use as hands. An engineer from another team walks up and shyly asks to take a photo. "Of course, yes!" he says.

There's a lot of mutual admiration in this oversized garage. The DARPA competition has brought some of the best and brightest minds in engineering from around the world under one roof, and the competitors seem just as interested in getting selfies with each other's robots as they are in taking first, second, and third place.

Many teams are pulling out extra cinder blocks, steering-wheel-shaped valves, and segments of drywall; since they're not allowed to see the course yet, they're building pieces of it to practice on in the coming days. Not so at JPL's stall. Kennedy shrugs, sipping a Pellegrino soda while RoboSimian stands silently behind him.

A few of the MIT students come over to greet Sisir Karumanchi, who used to work with them, and they hand him a T-shirt emblazoned with an ATLAS robot—their robotic contestant—riding a dragon.

"They dared me to wear this with the JPL hat," he said, smiling. "I told them I'd wear it once we beat them."

"Ladies and gentlemen . . . Start your robots!" Pratt says, revving up the crowd in the grandstands.

The first four robots are readied in their cars. They'll have an hour to make it to the end, and after a roughly half-hour break, the next four will come up to the stages. With twenty-four robots in the running, the six rounds will take all day. The competition is two days, so each team has two chances to get their highest scores possible. How they do on the first day will determine whether they're in the first or last round on the second.

Watching the robots "race" against the clock over the two-day race is both genuinely thrilling and utterly boring. Several whiz past orange barricades on the driving course, eliciting cheers from the spectators, but once they step through the door, Pratt's team hits their communication lines hard. Many a robot stands in front of the valve as the minutes stretch into infinity, as if pondering: to turn . . . or not to turn?

I'm trying to imagine what it was like for the engineers in the garage nearby, who can't see the course, and have to piece together what's happening in the arena with the tiniest dribbles of data from the robot. It must be nerve-wracking, like trying to take in a scene through a pinhole while someone is covering it with their hand, removing it only occasionally. And then, based on that tiny, narrow glimpse, as the darkness of the hand falls again, you have to plot your next move. It's moments like these where a robot's semi-autonomous software can really shine.

Back at the stages, the boredom is punctuated by the occasional dramatic collapse. One of the first robots of the day falls before it really starts the course. Another topples majestically and actually begins to "bleed out," spewing some kind of fluid. One robot falls so hard that its head pops off.

All of this is luridly fascinating, but in some ways the competition is just the tip of the iceberg. Outside the stadium, around the Fairplex grounds, scientists and engineers of all stripes have set up about seventy-five different booths where they're showing off their creations—small, cute-legged robots, exosuits, future space robots. There's even a tank meant for swimming robots (though I can't really tell how to view them, if they're in there) and an aviary for flying robots, where people cluster around a large net-covered tent. Boston Dynamics, the company that built the ATLAS robots being used by MIT and several other teams, shows off its animal-inspired legged robots, including a fan favorite: a four-legged robotic cheetah named "Spot."

Many of the "species" in this robotic menagerie might be used to aid human rescuers in disaster zones, or they might even be sent into space. You might not think of the interstellar void as a useful place for disaster robots. But a robot that can make its way over a rubble pile on earth might be able to climb a rock formation on Mars. The skills are surprisingly transferable.

This isn't a new idea. I first came across this concept in a flurry of papers in the late nineties and early aughts that sprang out of a 1998 NASA workshop at Jet Propulsion Laboratory on what were then called "biomorphic robots." I doubt this concept emerged in a vacuum: the workshop took place just a few years after NASA sent legged robots into active volcanoes (Dante in 1992 and Dante II in 1994) in part to understand how such limbed rovers could explore harsh terrain on distant worlds.

In these papers, the scientists and engineers proposed an "ecosystem" of crawling, swimming, and flying robots that would work in concert to gather data. These smaller, more agile machines would not replace the heavy, expensive, high-tech wheeled rovers that were being sent to Mars (and potentially other planets). Rather, they would complement them, examining hard-to-reach areas at low risk and low cost.

In one report, JPL researchers even presented an illustration: a rover or lander, parked near a steep cliff, while a swarm of flying robots gathers atmospheric readings and takes images, and a worm robot investigates some inaccessible nook in the distance. Such flying robots could be inspired by hummingbirds, bees, soaring birds like vultures, or even aerodynamic seeds that are shaped like wings so that the wind carries them. Surface robots could take after ants, snakes, or centipedes. Subsurface robots would look like jellyfish, earthworms, or germinating seeds—tiny, but powerful enough to work their way through packed soil, even rock. Small robots, acting in swarms, could work intelligently, almost like a superorganism (as bee hives and ant colonies are often described), assuming we learn to program them with the biological algorithms that define their behavior in nature. The NASA researchers estimated that a flock of 32 gliders, weighing just 75 grams each, could work together to cover a 10,000-square-kilometer area on another planet. They called such a program, aptly, BEES—Bioinspired Engineering of Exploration Systems.

The BEES researchers aimed to identify the principles in nature that allowed living things to do very basic, crucial functions with an ease that seemed to escape man-made mechanisms.

"The intent is not just to mimic operational mechanisms found in a specific biological organism but to imbibe the salient principles from a variety of diverse bioorganisms for the desired 'crucial function,'" Sarita Thakoor and colleagues wrote in one paper. "Thereby, we can build explorer systems that have specific capabilities endowed beyond nature, as they will possess a combination of the best nature-tested mechanisms for that particular function. The approach consists of selecting a crucial function, for example, flight or some selected aspects of flight, and develop an explorer that combines the principles of those specific attributes as seen in diverse flying species into one artificial entity. This will allow going beyond biology and achieving unprecedented capability and adaptability needed in encountering and exploring what is as yet unknown."

Such flying (and swarming) robots have yet to fully emerge from the scientific pupa; there are still a lot of kinks to work out, both in understanding flight dynamics at that scale and in programming in the kinds of behaviors that will allow robots to work cooperatively.

As I'm walking back from the robo-aviary, someone calls my name. My head whips around to an area where a crowd has gathered, and I walk over to greet Howie Choset, a Carnegie Mellon roboticist, with his face (wisely) covered in sunblock. Clearly he's been out in the sun all day, and isn't leaving any time soon. In front of him, several firefighters and other rescue personnel are standing in and around a little field full of cinder blocks and other debris while several robots wander around the enclosure. One firefighter grabs a little rover and hurls it across the arena; it bounces and easily rights itself upon landing. Another one demonstrates what appears to be an exoskeleton that seems to make holding heavy objects very easy. Each of these devices could potentially help rescuers in the field do their jobs more safely and effectively. Both kids and adults ooh and ahh at the various rovers.

But the crowd favorite might be the snake robot—Choset's addition to the crowded field—as it starts to climb up one of the men's legs.

You may wonder what a serpentine robot is doing hanging out with these other rovers, which, with their dark colors and treadlike wheels, seem like miniature tanks. In a conference where legs and wheels seem to be competing for best in show, the snake robot eschews both. But Choset's snake robot, and other robots like it, have talents of their own.

Like a lot of little kids, Choset was entranced by moving objects as a kid; wheeled things, like cars and trains, were among his favorites. That fascination with motion continued long into adulthood.

"I find freeway interchanges to be beautiful," he said. "Like where the 105 and the 110 intersect—that just blows my mind away."

After studying management and computer science at the University of Pennsylvania, he arrived at Caltech fully expecting to work on wheeled robots. But his adviser at the time, Joel Burdick, was building a snake robot with then-student Gregory Chirikjian, and Choset soon found himself hooked.

Since 1996, he's been working on snake robots at Carnegie Mellon, improving the design bit by bit. The wall of his lab is covered in snake robots, a sort of robo-serpentine hall of fame showing the snake's "evolution" over the past couple decades. He's made snakes that can slither, climb poles, and even perform heart surgery. He became particularly interested in the possibilities for search and rescue after meeting Robin Murphy, now a professor at Texas A&M University. He's taken the robot to visit her at Disaster City, the university's fifty-two-acre training facility featuring overturned trains, rubble piles, and even fire pits, used by teams of rescue workers and engineers to practice responding to a variety of disaster scenarios—earthquakes, fires, or even a nuclear contaminated hospital facility (complete with actors who play patients). At Disaster City, Choset would send his snakes into holes and crevices that a human body or arm could not go, or where perhaps there would be a risk of building collapse. Each time, the robot didn't work quite as well as he'd wanted.

"We would go, the robot would not work as well as we would like, we would get depressed," he said. "We'd go back, we'd fix the problems, we'd return, we'd overcome those problems, and we'd find new ones and we'd get depressed again . . . and the cycle repeats."

But Choset's snake finally hit a major stumbling block when he tagged along with Boston University archaeologist Kathryn Bard to investigate a site in Egypt in 2011. Bard was looking for secret caves filled with the skeletons of ships. The ancient Egyptians would sail the Red Sea to the ancient city of Punt, trade with them and then sail back north to Egypt. But when they landed, they didn't drag the ships back across the desert—they took the vessels apart and hid them in man-made caves, so that the next time they had to make the journey they would simply return to the hiding spot, reassemble them, and sail away.

Choset wanted to see if his robot could help her by squeezing into any inaccessible nooks. But whenever he tried to make the robot traverse a sandy slope, it flailed pathetically, getting nowhere.

"It was pretty sad," Choset said. "You want your robot to work well. We were hoping [to make] grand archaeological discoveries, and we didn't."

Choset's snake robot did, however, make quite an impression on Zahi Hawass, who was then still head of the Egyptian Supreme

Council of Antiquities, and a self-styled Egyptian Indiana Jones, who at times even wore a fedora rather like the sable one the whip-wielding archaeologist wore. (Though perhaps it's the other way around: George Lucas reportedly consulted with Hawass before creating the iconic swashbuckling character.)

While demonstrating the snake's capabilities, Choset had the robot climb up Hawass's leg. As Choset related to me, Hawass, known for having a bombastic personality, yelled, kicked the robot, and called for a photographer to come and take photos.

After things settled down, Hawass seemed genuinely enthusiastic about the robot. He sent the researchers out to the Giza Necropolis, to the causeway linking the Great Pyramid and the Sphinx, to explore an area known as the Osiris caverns. With only a few hours left in the day by the time they reached the spot, Choset's robot made it about fifteen meters into the cave, farther than anyone else before them—but probably still not far enough to reach the end of it. In any case, the whole trip left him hoping to come back and try again the following year. This would prove to be impossible: soon after they left, the Egyptian Revolution erupted in Tahrir Square.

Still, Choset took a valuable lesson from the experience. Until then, he hadn't really been studying the way that biological snakes move in order to program and build his robot. But having watched his robot struggle across the dunes in Egypt, he realized he might have to change tactics.

"There was only so far we could go with our intuition and innovation," he said.

A year or two later, Choset met Dan Goldman, a physicist at Georgia Tech, after hearing about Goldman's work on a nimble lizard known as the sandfish. This lizard lives in the Sahara desert and manages to move through its dunes with great ease; Goldman, a physicist, had studied how these animals moved and even built a robot to mimic their serpentinelike motion. This bot wasn't a prototype meant to be deployed in the field, though. For Goldman, it was a scientific tool meant to help him better understand the biophysics of the animals themselves—so he could slowly, as he improved upon the robot, home in on the fundamental physical rules governing their movement and define it in mathematical terms. It might seem that Goldman is studying terramechanics (the study of how vehicles interact with granular surfaces) but those classical laws were really meant for wheeled or tracked vehicles. Instead, he and others define this research as terradynamics, in an analogy with fluid and aerodynamics.

Sand isn't well defined either by fluid dynamics or solid mechanics. Individual sand grains are solid, but taken as a whole, they can flow, which makes them very tricky for wheels to navigate. Even legs can have some serious problems in this kind of medium, if they don't end in the right kind of feet, as anyone trying to navigate a powdery snowdrift without snowshoes will tell you.

Goldman has a bit of a thing for lizards. As a kid, he had his heart set on being a herpetologist (a scientist who studies reptiles and amphibians) but then veered toward mathematics, ending up with a Ph.D. in physics after a somewhat rocky road through graduate school. But he found himself doing some postdoctoral work in the lab of UC Berkeley professor Robert Full, whose PolyPEDAL lab has broken ground in understanding the physics of legged locomotion for decades. Goldman's choice to study granular physics also proved fortuitous, because a robot is only successful if it's in the right environment. A seal may swim but it sucks at sprinting; humans, even Olympians, are terrible swimmers compared to a dolphin. Understanding your medium is important, whether for evolutionary or engineering reasons, because the physical traits of that environment define the possibilities, and limitations, of what physical forms will work well. While we have robots that can travel on solid surfaces and robots that can swim and fly through fluids, we really don't know how to make robots that can effectively traverse sand, pebbles, and other granular media—as evidenced by the Mars rover *Spirit*'s untimely demise.

"We didn't have the fundamental physics equations of granular material," Goldman explained, "which is sort of surprising to people, that we can understand water, we can understand air, but we still don't have a full understanding of sand—in particular, sandy slopes."

There's one animal that seems to have a remarkably good rapport with sandy environments, and it doesn't even use legs to traverse them. That's the sidewinder, a rattlesnake in the deserts of the American Southwest that, if it had legs, would move something like a bellydancer moonwalking: sinuous, counterintuitive motion. This snake's movement doesn't match the direction it travels in—a phenomenon that boggles the brain, just as Michael Jackson first did when he seemed to step forward, but moved backward instead.

"There are field biologists who've studied these animals who say that if you look at sidewinding too long, you'll go mad," Goldman said, calling it "such a strange way to move."

Goldman was getting interested in the sidewinder's gait around the same time that he and Choset connected. Choset had also been trying to incorporate the snake's strange gait into his robot, and he had, arguably, the best snake robot in the world. The skill sets were totally complementary: Choset specialized in building field robots. Goldman's gifts, meanwhile, were part of his stock in trade as a physicist: the ability to take observations from the real world and distill them into elegant rules of movement. "He's very good at understanding in particular how mechanisms and artifacts interact with granular media . . . I'm just amazed at how he can see what we do and then draw it back into his field," Choset said.

In spite of their common purpose, Choset and Goldman have very different reasons for wanting to study snakes and build a robot. Choset's taken his robot around the states and across the globe to see how useful it can actually be, whether for human rescue efforts or archaeological digs. Goldman, on the other hand, wants to use the robot as a model to better understand the physics behind biology.

If you want to understand how a snake gets around, you have to test out all the different ways it can possibly move and then see which ones are the fastest, or most effective, or most energy efficient. Unfortunately, real live snakes don't take stage direction all that well. So a robot offers a model through which the scientists can understand the complex dynamics at play. And when Goldman uses the term "model," he isn't referencing toy trains—he's thinking about math. The robot is a physical manifestation of the mathematical formula that defines these animals' behavior. The scientists can tweak certain constants and insert different values until they finally hit upon the model that describes the snakes' reality.

The scientists usually use a known medium, like small glass beads or poppy seeds, to test their robots and animals, but they wanted to make sure that there wasn't something special about the sand's properties that defined the sidewinder's movements. So first author and graduate student Hamidreza Marvi and colleagues went out to the Yuma desert in Arizona, where the sidewinder *Crotalus cerastes* is found, dug up around a couple hundred pounds of sand, and trucked it across the country, all the way back to Georgia.

They took the sand to Zoo Atlanta, where they had built a small shed and where Joe Mendelson, the zoo's head of herpetological research, would bring over sidewinders and other snakes to test. Goldman had designed what he called a "fluidized bed"—a table filled with holes that used puffs of air to get the sand moving, giving it fluidlike properties and allowing him to erase the disturbances caused by the bed's previous serpentine test subject. (This happened right before each snake run; during the experiment itself, the bed was turned off.) The scientists had to cobble this little lab together in the zoo because many of the snakes were so venomous that there was no way they'd be allowed on the Georgia Tech campus.

After testing the sidewinders, they discovered what Goldman called a "beautiful" pattern: As the snakes ascended the sandy slopes, they never slipped. Their gait looked similar on a 20 degree slope as on a 0 degree slope, and about the same on hard ground, too.

If you watch a non-sidewinding snake move, you'll notice the undulating wave that it seems to generate as it moves, parallel to the ground. For the sidewinder, it was working with two waves of undulation—one that's horizontal (parallel to the ground) and one that's actually vertical (perpendicular to the ground). These waves happen simultaneously but they're not perfectly aligned: the vertical wave has a quarter-offset from the horizontal one, which gives the sidewinder its unique gait.

Unlike other snakes, sidewinders don't slide their entire bodies along the ground. Instead, only certain segments hit the ground at a given moment, and the snake uses this purchase to hurl the rest of its body forward, and then the newly advanced segments will now grip, throwing the trailing bits of the body forward again. I'm not sure entirely why, but this feels a little like walking to me.

"You got it. Sidewinding is kind of an interesting way to walk without legs . . . and then by phasing these waves appropriately, that's how you're walking," Goldman said. "So there really is a nice analogy that we haven't explored so much . . . and it would be fun to push on that."

Here's the thing, though: Choset's robot had already incorporated both horizontal and vertical waves into the robot. So why couldn't it traverse the slopes?

By analyzing the patterns left in the fluidized bed of sand, the researchers found that as the incline increased, the snakes were increasing the contact length of their body segments with the ground. On a 10 degree slope, about 40 percent of their body is in contact with the ground. With a 30 degree slope, their body contact jumps to 45 percent. The principle behind this is pretty simple—it's the same reason skis and snowshoes work in a snowdrift and high heels do not. A wider "footprint" allows you to distribute your weight, making it easier to maintain a grip without getting stuck—especially as the incline gets higher and the sand becomes more likely to collapse with even the slightest perturbation.

The researchers also tested thirteen other closely related pit vipers (under the watchful eye of Joe Mendelson, of course), expecting that perhaps some of them might be able to adopt the sidewinder's strategy when forced to travel up steeper sandy inclines. But the poor vipers thrashed and struggled, floundering pathetically.

"Most of them were pretty awful . . . the failures were just so surprising," Goldman said. "And we think that what's going on, and this is a hypothesis based on the robot, is they just don't have the right neuromechanical control schemes, even though it's a relatively simple scheme—just propagate these waves appropriately.

"They don't know how to coordinate their pushing and lifting, basically," he finished.

It seems that the snakes lacked the biological "software," even though it had the roughly the same natural "hardware" as the sidewinder. Which is an interesting point: Much of biologically inspired engineering focuses on the shape of the animal and how it interacts with its environment, as much of Goldman and Hu's work, separately and together, has previously done. But this project showed that understanding how an animal is programmed to move—how its software has evolved over millions of years—is just as important for designing future robots. A robot's software, not just its hardware, can benefit from bioinspiration, too.

Once they had turned the sidewinder's behavior into a formula and translated it into the snake robot's software, Choset's robot became "ridiculously maneuverable," Goldman said. If the roboticist ever returns to Egypt, he may have a little more luck.

Thanks to their collaboration, Choset now has a better robot, but for Goldman, the robot was never really the point. He was looking for a better understanding of the laws of nature—and the robot was the way to get there.

Robots were long considered an end in themselves; learning from the physiology of animals was simply a means to make them better. Now, that relationship cuts both ways: robots can now be a means to better understand biology by revealing the underlying physical principles at work.

Of course, that more intimate understanding of biology can be used to build even better robots, and so on. It's a virtuous cycle that causes Goldman to wax somewhat epic.

"If you're really trying to get crazy philosophical, we're trying to create life—I mean, living systems," Goldman said. "Living systems might one day not just be biology."

I wonder what Descartes would have thought.

The figurines posed on a high shelf in Robert Full's office look remarkably lifelike: A stalk-eyed crab, its claws brandished as if for a fight; a scorpion with its poison-tipped tail curled upward.

Farther down the shelf stand more cartoonish figures that are instantly recognizable for anyone who watched Pixar's animated film *A Bug's Life*—the movie's protagonist (an ant named Flik) and its villain (a grasshopper aptly named Hopper).

"Oh, I helped make them," the scientist said when I asked why the more naturalistic models were sharing space with animated characters. Full took forty hours' worth of footage of grasshopper faces, poking at the insects every so often to inspire a variety of facial expressions, and sent the most vivid ones to Pixar. He did the same for most of the movie's major characters, he added. "It was quite fun."

Full, a biomechanist at UC Berkeley, has been studying legged locomotion ever since he majored in biology at SUNY Buffalo, where he disproved an idea touted by established scientists that the energy cost of moving was based on leg number and published a paper on the findings during his senior year, a rarity for undergraduates.

"I published my first paper in *Science* and I thought, 'This is easy!' " he said, chuckling. As it turns out, he wasn't laughing at how easy it was, but at how naïve that feels now that he's a seasoned researcher with decades of experience.

Full might be best known for helping to solve the mystery of gecko adhesion—how those lithe little lizards manage to cling to smooth walls and ceilings with scant evidence of sticky glues and without so much as a toehold. In 2000 he and colleagues published a paper in *Nature* revealing the secrets of their gravity-defying grip: The soles of their ridged feet are covered in tiny hairlike structures called setae, and those hairs in turn sprout even smaller hairs called spatulae. A single seta measures about 110 micrometers long and just 4.2 micrometers wide—and there are more than 1.6 million of them on a gecko's front foot. All those tiny hairs vastly increase the surface area of the foot, and allow the animal to take advantage of a phenomenon known as the van der Waals force, an attraction between two atoms that only works on infinitesimally small scales.

That discovery helped spawn an entire new field and industry related to gecko adhesion, from creating gecko tape to building climbing robots. In fact, some of Brett Kennedy's colleagues at JPL are working on a cousin to RoboSimian (another version of LEMUR, in fact) that uses gecko-inspired grippers to slowly scale walls. One day, such robots could crawl along the outside of spacecraft, making repairs and sparing any astronauts the risk of performing a spacewalk and fixing it themselves.

Recently, Full has been turning heads for his research on that humble and yet apparently indestructible creature, the cockroach. On a nearby table strewn with scotch tape, push pins, and flyers for a new class on bioinspired design lies Rhex, a cockroach-inspired robot. Full and his colleagues have sent Rhex and his cousins into all kinds of places, running them through subways (to show that they could navigate urban environments and potentially prove useful in an emergency) and down the steps of Capitol Hill (to demonstrate to the gathered legislators the value of such science and engineering).

In spite of the potential applications for his robots, Full stresses that he's a biologist first and foremost.

"I do have a joint appointment in engineering but I don't see that as my goal," he said. "I want to understand nature."

Certainly, the terrariums full of docile cockroaches in the other room and the miniature stages mounted with high-speed cameras attest to that. Like Goldman's work with snake robots, Full's mechanical roaches are a natural result of trying to mathematically describe the dynamics at play in running, walking, and crawling animals.

"It turns out that it's hard to figure out what parameters to put in the math model. So it helps to have a physical model that meets the environment," he said. Full can change the stiffness of the leg or how his robot's feet work and the result, whether predicted or surprising, informs how he tweaks the mathematical model.

If there were such a thing as the pinnacle of evolution, it would likely be the cockroach, not *Homo sapiens,* perched at the very top. Anyone who has had a cockroach infestation knows how devilishly difficult they are to catch and to kill. Recent research has shown how quickly the insects have evolved to avoid the sugary snares in deadly traps. These insects can go without water for weeks and even survive decapitation for a limited time. If there's anything that can survive a nuclear disaster (or other global catastrophe), it will be these hardy little insects. What better insect, then, on which to model a robot?

Full has long been fascinated by how these creatures manage to maintain their stability over uneven terrain. He and his colleagues have strapped tiny jetpacks and shot them off like pellets, or had them run across plates that they then jerked around. In both cases, the insects found their footing with remarkable speed—too fast, in fact, for it to be explained by a signal traveling to a neuron in its tiny brain and back. He's discovered that a lot of their remarkable abilities have to do not with their fast reflexes, but with something a colleague named "preflexes"—behaviors that are built into the very materials comprising the roach's body.

"Instead of neural feedback, we call it mechanical feedback," he said.

After studying live cockroaches, the researchers carefully tuned the robot's material properties. They've also applied the same principle to building an exoskeleton made of overlapping plates that deforms the way that a cockroach's exoskeleton would, allowing it to be tough enough to protect its innards but flexible enough to squeeze into surprisingly tight spots.

This armored robot, called CRAM, can spread its legs and adjust its gait to maintain speed even in cramped quarters—a trick the scientists learned from watching real cockroaches squeeze through tunnels that were merely 4 millimeters high, less than a third of their standing body height. They used a force-measuring device to "squish" cockroaches and found that they could be flattened by up to nine hundred times their own body weight and spring back as if nothing had happened. (No cockroaches were actually harmed in this experiment, for those worried about the insects' well-being; they all flew and walked normally afterward.)

In the tightest of spaces, when the insect can sprawl no more, it uses a strange gait that the scientists called body-friction legged crawling—a movement that might not be too far off from the way that animals such as snakes move across the ground, which is what Dan Goldman is working on at Georgia Tech. (Goldman also happens to have been Full's postdoc years before—heightening my sense that the robotics community, while a worldwide research-industrial complex, is in some ways surprisingly small.)

Full expects that this kind of research will help scientists think differently about the kinds of animals that can inspire soft robotics; it's not all worms and jellyfish. The hybrid materials of the insect exoskeleton might prove essential for roboticists, given how seamlessly they incorporate hard and soft elements.

Out of curiosity, I ask him what he thought about the DARPA challenge, held a few months before. Full took some issue with the idea that such rescue robots required human form. "The logic breaks down at some level," he said. After all, just because a human paints walls doesn't mean that an insect-shaped robot couldn't do it; they simply might do it differently.

"I think it demonstrated that we weren't ready for doing humanoids—that was my advice," Full said of the challenge. "What it did demonstrate was how hard it is to do robotics."

4

HOW FLYING AND SWIMMING ANIMALS GO WITH THE FLOW

Geoffrey Spedding's office sits inside the Olin Hall of Engineering on the University of Southern California campus, not far from a statue of USC alum and NASA astronaut Neil Armstrong, which I pass on my way there. The décor seems typical for an aeronautical engineering professor. A wooden model glider lies across a wide filing cabinet, its wings and body neatly disassembled; on a wall behind his desk, he's pinned up a photo of a sleek spy plane as well as printouts of line graphs showing the results from his favorite aerodynamics experiments.

But next to images of fighter planes hang a handful of incongruous items: a photo of a tiny, translucent sea creature known as a copepod; a photo of a long-whiskered seal, with large, anime-sized eyes; and a single, ombre-gray seagull feather.

Spedding is a professor in and chairman of USC's department of aerospace and mechanical engineering, but his doctorate is actually in zoology, and the interest shows. The scientist has worked with Swedish researchers to put birds and bats in wind tunnels to examine their aerodynamics. And when I first met him in 2010 at the American Physical Society's Division of Fluid Dynamics meeting in Long Beach, around the time he'd become department chair, he told me that he was planning on introducing more bioinspiration as a major emphasis in the department.

"Don't write that down yet, not everybody knows that," he said at the time. There was a reason for his caution; for at least two decades, it had been hard to attract engineering graduate students with an emphasis in what he called "bio-fluids," because of fears that they couldn't be guaranteed a job after graduation.

But that attitude has been steadily changing, he added. Spedding used to regularly plot the rising number of bio-fluid mechanics sessions at these conferences versus the number of traditional turbulence sessions, "to convince people I was doing something that had a future."

"There's been a dramatic cultural change," he added. "Gradually people realize that there are really interesting problems that involve the overlap between biology and engineering."

Now, more than five years later, his interest is out in the open. I ask him about the seal and he explains, with great enthusiasm, what it has to do with his work.

Spedding doesn't just study how things move through air; he also studies how they move through water and more to the point, the trails they leave behind. You might think, as a fish swims, it leaves no physical evidence of its presence. It's not like it can leave footprints or fingerprints on a solid surface, after all. But that's not quite true: a fish (or anything else) moving through the water actually produces a complex wake, a pattern of vortices and eddies that lasts much longer than you might expect.

Spedding and his colleagues have performed simple exercises to demonstrate this phenomenon, moving a sphere through a "perfect" ocean that's free of confounding disturbances, and found that these telltale patterns could last for a full ten days. Granted, the actual ocean is not perfect, but even in a messy, sloshy body of water, the scientist guesses that a wake from a large-enough moving object could probably last a full day.

The U.S. Navy, for example, pays attention to this kind of research. After all, what's the point of a submarine, swimming stealthily beneath the water, if you can tell where it is and where it's been as easily as if it were a boat skimming the surface? On the flip side, it would also be extremely useful to know how to track other nations' submarines.

All sorts of people appear to take interest in his work, Spedding says, including folks he's never met.

"They'll sort of tap me on the shoulder and say, 'How about other shapes?' without specifying what the other shapes should be," he said with a laugh. "So we did other shapes, and I gave a briefing in Washington D.C. Some strange people came to the briefing who I'd never seen before; they came in the back and they left early, and then they trapped me in the stairwell on the way out of the seminar. I'm being a little overdramatic, but not much. They kind of found me in the stairwell as I was going down after the seminar. They said, 'We really liked parts of your seminar. We can't tell you which parts, but we really liked those parts.' "

Spedding could learn which parts they're interested in, if he wanted to—but such knowledge would come with a price, one he's not willing to pay.

"I've been asked on a couple of occasions if I want to have security clearance and I say no, because that way I can think whatever I want to think and say whatever I want to say, because nobody told me," he says. "It's much more convenient that way. I can just bumble around and maybe I'm on target, maybe not, but the reasons I do the things I do remain just because I think the problem is interesting or difficult . . . and then I don't have to worry about that. They can worry about that stuff."

The potential to track any swimming object isn't purely theoretical. Animals seem to do it all the time. Take the harbor seal pinned to his office wall; those limpid eyes and improbably long whiskers aren't just for show.

"They live in murky water, which is why they have these cute, huge eyes," Spedding points out. "But sometimes, the water is so murky that eyes are useless, and that's why they have these huge whiskers."

There were a series of experiments published in the journal *Science* where researchers trained a seal to follow a toy submarine around in a swimming pool and rewarded it with a fish when the seal managed to find it. The scientists would put a stocking over the seal's head, bank-robber style, so that it couldn't see, but they'd leave holes for the whiskers to poke out. They'd put headphones over its ear holes to block out any sound, and then let the little submarine chug across the pool, until it came to a full and complete stop. The researchers would then remove the headphones and let the still-blinded seal go searching. Most times, the seal would easily find the submarine—not by going directly toward the toy, but by tracing the same path it traveled in the pool, as if following an invisible wake.

Those sensitive whiskers, Spedding thinks, might be able to track the pattern of vortices and eddies left by swimming things in a way that human technology still can't. It makes sense, Spedding said—there are documented blind seals living in the wild, which means they must be able to track and catch prey well enough to survive. (Another detail bolstering that theory: When the stocking also covered the whiskers, the seals were totally at a loss.)

"That seal must think humans are crazy," I can't help but interject.

"I know," he says. "And it's right."

The seal isn't the only animal that might be taking advantage of this phenomenon. One of Spedding's Swedish collaborators on bird flight was married to a scientist, Susanne Åkesson, who studies sea turtle navigation. The green sea turtle starts life off in a very small number of places—one of them being Ascension Island, a tiny speck of rock in the middle of the Atlantic Ocean. Once they hatch in the sand, those that survive the harrowing journey through the gauntlet of hungry birds and violent waves swim to far-off locales—typically, somewhere along the eastern coast of South America. Here's the crazy part: When the females are mature and ready to lay eggs, they make the arduous journey back to that exact same tiny island where they were born. They can do this, regardless of where they ended up. How is this possible? Spedding speculates that it may have to do with the wake of the island itself, that it leaves a telltale signature in the ocean's thermocline layer, which is deep enough that it's relatively unsullied by wind, animal motions, and other disturbances. The turtles, by executing regular deep dives (which make dramatic, thin V-shapes on the graph Spedding shows me) may be able to pick up and follow this incredibly narrow trail, etched out by a tiny island in a vast, overwhelming ocean.

When I first met Spedding at that fluid dynamics conference, I had expected the event to be kind of dull. I soon found out how wrong I was. Alongside presentations concerning the atmospheric dynamics of Jupiter and on the behavior of milk poured into tea, I met a host of scientists working on problems related to how living things, whether flying or swimming, move through a fluid world. One presentation I saw even analyzed how ants acted like fluids themselves—able to climb up a glass wall as if they were part of some thick, sticky liquid.

Until this meeting, I hadn't realized that birds and fish are essentially doing the same thing: moving through a fluid medium.

And as I spoke with researchers it quickly became clear that animals (and some plants, whose wing-shaped seeds can travel for miles) seem to grasp fluid dynamics' secrets in a way that human engineers simply did not.

The research that Spedding, one of the conference's organizers, presented in a session on flight served as a perfect example: why don't planes look more like birds?

"Of course planes look like birds!" I thought to myself. They may not flap, but they do have beaky faces, wide wings, and a tail, after all. Many flying vehicles, including hot-air balloons and helicopters, lack such apparently avian appendages.

But the truth is that engineers have forgotten many of nature's lessons since Wilbur and Orville Wright drew inspiration from the birds they observed. At the time that the bicycle-fixing brothers were working on their prototype, fellow aspiring aviators were running into serious technical holdups. Other engineers' craft had long, stiff wings that provided stability but robbed the pilot of much control. Many died in attempts to perfect their craft—including the German engineer Otto Lilienthal, nicknamed the "Father of Flight" for his contributions to the field (and for popularizing the crazy notion that these gawky contraptions would one day become practical modes of global transportation).

Before he died, Lilienthal had begun to believe that birds held the key to true flight, hypothesizing that a flying machine would require more birdlike features—a flapping wing, to be precise. The Wrights did not take to that idea, and it's good that they didn't; flapping wings are extremely energetically costly unless you're actually bird-sized. But the brothers did develop their own bird-inspired innovation: a warping wing, which could change its shape on command to give its pilot unprecedented control when banking right or left.

Since the brothers' successful flight near Kitty Hawk in North

Carolina, planes have radically diverged from the bioinspired flight path. Today's plane is not all that aerodynamically designed—the modern 747, for example, has a long, unwieldy body that produces far more drag than wings alone would. This is, of course, for a host of logistical and technical reasons: today's passenger jets typically breeze through the clouds around five hundred fifty miles per hour, while birds generally clock in an order of magnitude lower, perhaps around twenty to thirty miles per hour. Still, many of the decisions made in modern airplane design, Spedding argues, are made simply because that's the way it's always been done, not because it really is the optimal design.

In his office, the scientist pulls out a sheet of paper that resembles a sort of airplane family tree. He points to what looks like a thin crescent moon at the bottom—an idealized wing, as seen from directly above. This would be the most aerodynamic shape. The problem is, you can't really fit passengers or cargo inside of a skinny wing (unless that cargo is liquid, like fuel), so you need a proper body. But once you stick a fuselage in the middle of those wings, that giant obstruction starts to reduce the lift and raise the drag.

Spedding moves his finger to the next iteration: a "flying wing," slightly thicker and fatter than the ideal wing, perhaps most infamously built by two Nazis, the Horten brothers, near the end of World War II. Another branch leads to a design called the "blended wing body," a concept that Boeing has been mulling for several decades to carry passengers but has never really built. (It's test-flown some unmanned versions in recent years.) The rest of the nodes on this irregular tree look somewhat like the plane you're familiar with—long, cigarlike fuselage, wide, straight wings, and a long tail with mini-wings at the end.

"This is the configuration that we use today," Spedding says, pointing at the sketch of something that looks a lot like a jumbo jet. "But this cannot be the most efficient solution because aerodynamically, this sucks. It sucks because the aerodynamics of the wing cannot be maintained over the fuselage, and so what started off as a nice clean aerodynamic solution here is disrupted by the presence of this fuselage."

The idea that perhaps airplanes were imperfectly designed—and that there could be a better template—came from Spedding's coauthor and collaborator, Joachim Huyssen, a Ph.D. student and aeronautics engineer at the University of Pretoria, in South Africa. Huyssen remembers vacationing by the ocean in the summer, watching swifts and swallows perform acrobatic somersaults in the strong coastal winds. He learned how to hang-glide in college, and on a trip to Germany watched young children on a bridge in Salzburg feeding bread to seagulls, who were able to easily swoop in and nab the scrap out of little fingers without actually flapping their wings.

"In a way I've architected my life around aeronautical engineering," he tells me over the phone from his office at the University of Pretoria, which just happens to be Spedding's former digs. "It has always fascinated me . . . why do birds and airplanes look different?"

Before we go further, it's probably worth setting up a quick primer on planes and how they move. The main concern of a flying vehicle or animal is the force of gravity, which wants to drag you down to earth. To counteract gravity, you have to produce upward force, or lift—which can be generated by a wing as it moves forward. But that forward motion causes pushback from drag (of which there are two main types: pressure drag, the dominant type, caused by the resistance of all the air molecules as you try to push forward through the air; and friction drag, caused by the air rubbing against the flying object's surface). So to counteract drag, you have to produce forward motion, thrust (using wings for a bird, or engines for a plane). These four forces, in four different directions, define the movement of living and man-made fliers alike.

Whether you're a real bird or piloting a giant metal one, you basically want to generate as much lift as you need (keeping you off the ground) while minimizing the drag, so you don't have to waste a lot of energy to keep the thrust high. The higher the lift-to-drag ratio, the more efficient the flier. But how exactly do you generate lift?

There are many explanations for what causes lift, and as many explainers arguing that the others are wrong. I have watched and read many lectures by learned professors complaining about incorrect explanations—only for the next lecture to contradict the previous one! For the purposes of keeping things somewhat simple, I'll try to explain here how lift works without going into all of the problematic interpretations, and hopefully without offending anyone in the process.

For a relatively simple explanation, we'll turn to one of the simplest airfoils you can imagine: A curved, flat plate, which Spedding draws as a line on a piece of paper. Both the air that travels over the top and the air that passes under the bottom are deflected downward as they reach the airfoil's trailing edge. According to Newton's third law of motion, for every action, there is an equal and opposite reaction, and because the airfoil pushes all that air down, the airfoil itself must move up.

If, for whatever reason, the lift generated by the plate isn't enough to keep it aloft, you can tilt its nose toward the sky. The more you raise this "angle of attack," the more air it throws down, and the more lift it generates.

But there's a problem with that strategy: all that tilting starts to mess with the boundary layer, a thin layer of air that hugs the top and bottom surfaces of the wing and that actually takes on special properties of its own, distinct from the bulk air around it. If you've ever driven upward of eighty miles per hour on an empty highway and wondered why water droplets on your car don't seem to be affected by all the wind rushing by, it's because they're riding in that special boundary layer.

Ideally, the boundary layer will stay attached all the way from the front of the wing until it passes the end of it. (Though I will admit that this is probably unscientific and inaccurate, I tend to think of it as kind of like having good traction on your shoes while running.) The problem is, as you raise the angle of attack, some of that "traction" starts to wear off the back edge—the boundary layer will start to separate earlier and earlier. The more that nose tilts upward, the more that separation point creeps toward the front of the wing. Because of all the lift that's being generated, that separation isn't much of a problem—until you get to the "critical angle of attack," when the nose tips so high that the boundary layer simply can't stay attached at all and completely lifts off of the surface. That allows stagnant air to rush in and fill the void, raising the air pressure on top of the wing and causing the drag to spike. When that boundary layer separates, the wing is no longer hurling any air down—which means there's essentially nothing keeping the whole contraption up.

The critical angle of attack is different for different kinds of airfoils, but its effects remain the same: The plane will suddenly falter, even though the engine is still running. Pilots know that sickening feeling well, and they generally do their best to avoid it, because the plane can go into a tailspin if it isn't expertly brought back under control. You can see this most clearly if you launch a paper plane at a really steep angle—it will climb upward but then suddenly pause before taking a nosedive. That pause, right there, is stall in action. So airfoils need that boundary layer, that conveyor belt of air with special properties, to stay as close to the surface as possible for as long as possible.

Thus, airplanes must be able to control their pitch, or tilt of the nose. And it's actually kind of hard to do, because as air goes over the wings, the plane naturally wants to nose up or down. So engineers added the tailplane—a second pair of miniature wings, to stabilize the aircraft. Then they lengthened the fuselage to make that tailplane more effective. The result? A profile for your prototypical passenger jet.

The tailplane wasn't always a part of aircraft: the Wright brothers, for example, developed a pitch-controlling device that actually jutted out in front of the airplane. The brothers' realization that they needed not just to generate lift, but to actually control flight in all three dimensions (pitch, roll, and yaw) was in large part why they succeeded where so many other aspiring aviators failed, Spedding said.

The main thing that bothered Huyssen about the modern pitch-control mechanism was this: if these tails are so darn useful, then why don't birds have them?

More than two decades ago, Huyssen dedicated his research at the University of Pretoria to trying to design a glider from first principles, building a radio-controlled model that ended up looking remarkably birdlike, with gull-like wings and virtually no tail. The idea was that instead of controlling the aircraft with a tail (or with flaps on the wings the way conventional aircraft do), this aircraft would control its movement by the back-and-forth sweep of its bent wings, the way a bird might. The engineer was so confident that in 1995, he even harnessed himself into a full-scale glider and had it lifted high into the air by a hot-air balloon to be launched.

Video recorded from the balloon shows where this first (and

final) free-flight test went horribly wrong. Moments after the glider is released, it suddenly snaps into pieces and spins wildly out of control. In the footage, a man hanging from the balloon cries out as he watches the fragments careen away. It's kind of horrifying to watch, especially since it's not clear from this vantage point what happens to Huyssen.

The engineer fortunately had a parachute that deployed, bringing him safely back to the surface. But in a terrible twist, the glider's collapse was probably caused by that very same parachute; the seller had sent one that was much bigger than what Huyssen had ordered, and the glider apparently could not handle the additional mass. With little funding to repair the aircraft and continue his studies, the engineer was forced to abandon his plans, as well as his Ph.D. Perhaps, like Icarus, he flew too high, too fast.

Huyssen started an aeronautical engineering company and maintained a workshop on campus, but he didn't return to his research until 2009, after hearing about Spedding and his work in animal flight. Spedding was doing a brief, year-long teaching stint at the university, because his wife is from the area, and the younger engineer sought him out during that stay.

"He explained the idea to me and I said, 'Wow, that sounds really interesting,'" Spedding recalled. "And he said, 'Do you think this is right?' And I said, 'I have no idea—but it doesn't sound wrong in principle.'"

Spedding returned to Los Angeles but he stayed in touch with the South African engineer, realizing that in order to solve this mystery, he'd have to bring together three things: Huyssen's brain, his models, and USC's wind tunnel.

Eventually, the professor secured a grant that allowed him to bring Huyssen over for just a few weeks so that they could test the model and measure the lift. They tried three different configurations: a simple flying wing, then the wings with a body, and finally the wings with the body and a stubby, birdlike tail.

"It was totally frantic towards the end," Spedding said. "We had about two weeks to set up, one week of fiddling around and trying to get things working properly. And so in fact we got all of the data in just a few days. It was really—we didn't know if we were going to make it."

Even when you do such experiments, you often don't know immediately how well they worked. The flows of air over a body are incredibly complex, and it takes weeks to properly sew all the packets of digital data together. But once they did, it became clear that the stubby, birdlike tail actually was working the way they had expected it to: grabbing the boundary layer that was separating off the back of the body, forcing it to follow the tail's surface, and thus restoring the lift over the body as if it were going over the simple flying wing.

However, you might point out, just because it's more efficient doesn't mean it's the best design to carry people and luggage, as well as feature well-placed lavatories and emergency exits. But the researchers have you covered. Huyssen and his students originally showed that the packing efficiency of people and their luggage was actually better with these stubby silhouettes than in the standard airplane body. A couple years later, Spedding and Ilaria Mazzoleni, a professor at the Southern California Institute of Architecture, ran a two-week workshop where the students produced an array of designs that envisioned what the interior of such a bird-shaped passenger plane might look like.

The researchers haven't yet quantified exactly how much better this plane might be, lift-to-drag wise, compared to a typical airplane. Part of the problem is that air acts differently around something as small as a seagull than it does around something as large and fast as an airplane. The proof would be in actually building a human-sized model and testing it out—perhaps even taking it to compete in notoriously competitive glider competitions. But that takes time, people, and money. As we walk over to check out the wind tunnel, Spedding hazards a guess: $100,000, two dedicated engineers, for two years.

In the meantime, Spedding and his students continue to delve into the details of how air flows over wings, finding some pretty surprising phenomena along the way. At the door of a giant, windowless metal building, the scientist plugs in a keycode and we go in. We sidle past a massive low-density gas chamber that researchers use to test the effect of tiny forces exerted in space, past sheets of carbon fiber used by the rocket lab next door, and open a door labeled FLIGHT EXPERIMENTS with a warning sign: DANGER LASER OPERATING. An electric sign that reads LASER IN USE is unlit; right now, it's safe to walk inside.

A good wind tunnel is hard to find—which is why, even when they retire from their government jobs, they might find work in the private sector. The Dryden wind tunnel served the National Bureau of Standards in Washington D.C. since the 1930s and was later disassembled and sent over to the University of Southern California in the 1970s. Spedding proudly calls it one of the best wind tunnels in North America for low-speed flows, and he still regularly puts the venerable old device through its paces.

The wind tunnel, from my vantage point, looks huge—extending back what looks like fifty feet—but the heart of it is a pretty modest affair: a wooden tube, four feet across, that I could probably stand in if I stooped a little bit. Mounted inside, visible through a glass window, is what looks like a sheet, folded lengthwise to produce the teardrop shape of a prototypical, symmetrical wing. It's mounted vertically, as if flying on its side, and for at least twenty minutes I didn't even realize that the tunnel was actually running—the air rushing by the wing is invisible, and the low hum from the machine is pretty discreet. The machine occasionally emitted a loud buzz as it repositioned the wing at a different angle of attack, in order to measure how the forces change at different angles; a handful of undergraduates and Ph.D. student Joseph Tank watched as new data points popped onto a line graph on the computer screen in front of them.

Tank is studying a wing that's just a few inches long—in the range for a small bird. That's no accident, Spedding says; they're interested in this regime for the sake of understanding the aerodynamics of birds as well as small, fixed-wing drones. The graduate student is plotting the amount of lift generated by this wing against the angle of attack. As we've learned before, raising the angle of attack from flat (0 degrees) should also raise the lift (until you get to that critical angle where the flow separates). It's a well-known relationship, and it results in a very straightforward graph, a straight line (until that critical angle).

Or at least, it should. But there's something very, very weird about how this particular size range of wing affects the air at this particular wind speed. This graph, which should be a straight line, actually has a squiggle in the middle that looks like the mirror image of a letter S rotated 45 degrees.

"Look at this!" Spedding says. "This is completely crazy. Absolutely mad."

He points to the two deviations from the straight line, one dipping below it and the other rising above it.

"This is the straight line that it ought to be," he says, pointing to the slanted line superimposed over the squiggly data. "This is the way every drone you've ever seen models the lift and drag on its propellers; this is the way every flight model you've ever seen, with birds and so on, models the lift versus alpha. But it doesn't work."

As you go from an angle of attack of zero (perfectly flat), your lift should increase with no interruptions, following that straight line that Spedding pointed out. But Tank's data seems to show that, as you go from zero angle to a slight angle, the lift actually decreases before recovering and then, somehow, exceeds the expectations set by that straight, diagonal line. Spedding's of course impressed by the much-better lift at higher angles of attack (or "alpha"), but it's that strange dip in the beginning that has him hooked.

These are the types of experiments that Spedding is most interested in: deceptively simple, basic exercises that give confounding results that others have either missed or ignored. For most engineers, designing their wings based on the straight-line graph might be a good-enough approximation, but it won't get you to an optimally designed aircraft. In some ways, it's akin to the thinking Lilienthal and others exhibited when they tried to make planes that looked feathered, the way birds did, without a real understanding of the dynamics at play. (To be fair, Spedding said, part of the reason is simply that engineers had no reason to study tiny wings until flying drones became a reality.)

The only way to build the best flying machine, bioinspired or not, is to delve into those devilish details to find the physical principles that may lie underneath.

Science is not easy, though. Tank pulls up a graph of the current experiment's drag curves, and they don't look nearly as clean as the squiggly lift graph—Tank thinks it has to do with a change in his calibration method. It also turns out that a new force-measuring stand they had just installed appears to have a crooked screw, and so he'll have to take it apart and adjust it to use the instrument properly. Spedding leaves his graduate student to keep collecting the confounding data, but not before asking for a print-out of the lift-to-alpha graph. He wants to add it to the wall of favorites in his office.

Certain people seem fated to walk particular paths in life—and I'm reminded of that every time I talk to the aptly named Frank Fish, who runs the equally alliterative Liquid Life lab at West Chester University in Pennsylvania. The biomechanist didn't really set out to study fish and other swimming animals, but his research seems to have evolved in that general direction over time, starting with muskrats to seals and then eventually dolphins.

"I guess it sort of just came," he says.

Dolphins, with their smiling faces, big brains, and major megafaunal charisma, are perhaps best known by humans for their intelligence and playfulness. But not enough attention is paid to their impressive swimming abilities—while surfing, I've actually watched dolphins ride an overhead wave in choppy waters off the coast of Cape Canaveral that none of the shortboarders around me dared take. Perhaps dolphins' prowess is easy to take for granted—they live in the ocean, after all—but exactly how they swim so well has actually baffled scientists for decades.

Much of the trouble can be traced back to British zoologist James Gray, who in the 1930s wrote about a report by a man named E. F. Thompson, who while on a ship in the middle of the Indian Ocean watched a dolphin swim so fast that it passed up his vessel in just under seven seconds, "as timed by a stopwatch," Gray wrote. Based on the speed and length of the ship (8.5 knots, 136 feet), he calculated that the dolphin must have been swimming at 20 knots, or 33 feet per second. This should be impossible, Gray mused; according to his calculations, dolphins simply didn't have enough muscle power to swim with such great speed, because the drag as their bodies moved through the water should have been holding them back.

"If the resistance of an actively swimming dolphin is equal to that of a rigid model towed at the same speed," he wrote, "the muscles must be capable of generating energy at a rate at least seven times greater than that of other types of mammalian muscle."

If that sounds nuts, Gray seemed to think so, too. The scientist concluded that there must be something else special about the dolphin's body that allowed it to overcome the drag and move through the water with comparatively less muscle power. This became known as Gray's paradox.

One of the favored solutions to Gray's dilemma was this: That there must be something about the dolphin's smooth, gray skin that allows the animal to reduce the drag on its body. Over the years, many researchers tried to discover what these mysterious properties might be, and the pursuit of this material took on new urgency during the Cold War era, after German aeronautics engineer Max Kramer published research in 1960 claiming that he had created a faux torpedo wrapped in an artificial skin that mimicked the dolphin's. This skin, which featured a viscous liquid layer underneath the rubber, reportedly reduced the drag by 59 percent. This played right into the fever dreams of both the Soviet Union and the United States, drawn by the lure of coating submarines, ships, and actual torpedoes in this skin to help them go faster. Each nation believed it would give them the edge on the other (whose submarines, it was suspected, were much speedier than their own).

This avenue turned out to be a dead end. Many tried to reproduce Kramer's results, with no success.

"When the Americans started looking at Kramer's work, they couldn't replicate it," Fish said.

Researchers then wondered if the skin folds that seemed to ripple across the dolphin's body were somehow smoothing out the flow. To see if this was the case, in 1977 Russian scientists went so far as to drag naked young women between the ages of seventeen and thirty years old with a tow line to see if any ripples traveling across *their* skin helped reduce drag. The logic was this: Women (particularly young women) tended to have more adipose tissue, or subcutaneous fat, underneath their skin, making it more compliant and dolphinlike. Of course, the experiment revealed that the rippling skin folds increased drag; wearing a swimsuit actually helped mitigate it. Pictures of the "nekkid leddies" that were included with the work "indicate visually the beauty and excitement of experimentation," one almost-certainly-male book reviewer wrote.

Fish and his colleagues decided to put this question to rest using a modern technique to more accurately measure how dolphins moved. The technique, known as particle image velocimetry, actually takes a tank full of water, fills it with tiny, metal-coated glass beads and sends a laser sheet through, illuminating the suspended beads. When an animal swims through the water, Fish says, the beads move through the water with the flow, revealing its complex dynamics, and a high-speed camera captures the laser-lit action for analysis later.

But dolphins hold a special place in human society, even among researchers, and Fish knew he wouldn't get approval to fill a dolphin tank with beads. They're large, lovable mammals, and there were legitimate worries that they might swallow the tiny beads or the lasers might hurt their eyes. Luckily, Fish had heard about an ingenious (and inexpensive) method being used by Timothy Wei of the University of Nebraska–Lincoln to study the movement of another such "pampered animal," as he called them: human Olympic swimmers. Instead of using glass beads and a laser sheet, the scientists made a screen of tiny air bubbles using a garden soaker hose that pumped in oxygen from a tank. The bubbles were so small that they wouldn't pop, and trained dolphins could easily swim through them and generate the same effect.

With the help of two former U.S. Navy personnel—bottlenose dolphins Primo and Puka, now retired to UC Santa Cruz—the scientists were able to track the bubbles with a high-speed camera just as they would the beads. Then they could look at the bubbles' velocities, the path they traced, the vortices they formed, and with the help of a computer algorithm they could use that information to calculate the animals' thrust.

As it turned out, the premise for Gray's paradox was wrong, on a couple of levels. The dolphin that he used as the basis for his analysis may not have been traveling at his calculated 10 meters per second; Fish was only able to get the animals to swim at 3.5 meters per second, which, while it may not have been their top speed, still seemed to be a good clip. (The fastest he'd ever seen them go was 11 or 12 meters per second, but that was only as they were accelerating for one or two seconds upward as they prepared to leap out of the water and perform an aerial maneuver. But jumping out of the water is very different than sustaining that pace to pass a sailboat, and even requires very different muscle fibers; just try and ask a sprinter to run a marathon at top speed.) Fish also suspects that the dolphin in question may have been traveling close enough to the boat and thus was able to take advantage of the wake, artificially speeding up its progress.

More to the point, Primo and Puka revealed that the dolphins really were producing far more thrust than Gray had expected, and enough to power them as they sped through the water. In other words, they didn't need any help from supposedly superpowered skin.

"The real paradox was that, despite its inaccuracies, Gray's paper was the impetus for novel innovations that have furthered the areas of dolphin biology, hydrodynamics and biomimetics," Fish wrote in a paper for the *Journal of Experimental Biology* summarizing his conclusions. "There is still more to be discovered; perhaps the dolphin has not given up all its secrets."

If Fish seems overly forgiving of all the trouble Gray caused, perhaps it's because he knows what it feels like to have your most basic assumptions overturned.

Some three decades ago, when Fish had just started dating his now-wife, the pair were strolling around Quincy Market, a touristy area in Boston. One particular shop seemed full of animal art, perhaps inspired by the marine life just a few blocks away, both in the nearby New England Aquarium and the ocean itself. Among the tchotchkes, a particular piece caught his eye: a little figurine of a humpback whale with its flipper raised. To the biologist, something about the flipper seemed deeply off.

"I looked at it and I go, it's wrong. It has these bumps on the leading edge," Fish said. "I know fluid dynamics. You don't put bumps on a leading edge—look at an airplane. It has a nice straight edge."

Fish informed the shopkeeper of the error. But Bostonians are not known to be shy with their opinions, and the shopkeeper told the biologist in no uncertain terms that in fact he was mistaken, and the artist had gotten it right.

"I got rather obsessed with this," Fish admitted. "I wanted to know why I was wrong."

Fish decided he'd try to model and study this paradoxical appendage and called a friend of his at the Smithsonian, asking if he happened to have a humpback whale flipper lying around. At the time, none were available—but the friend promised he'd put Fish's name down if any flippers came in.

Months passed. Then one day, Fish's secretary surprised him with a strange message.

"Someone called—they said there's a dead whale in New Jersey and you're to come and get it," she said.

"What?" Fish said.

Apparently, this was part of the deal. Scientists can't harvest live whale parts, for obvious ethical reasons. They have to wait until a beached whale dies and (assuming there was no way to save it) then scientists can take the body away and store it for future research. So if Fish wanted a whale flipper, he would have to go and grab it himself. Fish quickly called the folks handling the carcass. How big was the whale? Twenty feet, they said. The biologist did some quick mental math: That meant that the flipper was probably about six feet, about the right size to fit in his compact Mercury Lynx. He drove up to the New Jersey Marine Mammal Stranding Center to pick up his smelly trophy, only to find, when he got there, that the flipper was a whopping ten feet long. Fish had to cut the flipper into three sections and stuff the pieces into the trunk of his car. The weight was so great that it forced the rear of the car to sink down almost to the ground.

"I was afraid that as I drove home, a New Jersey state trooper would stop me and ask what was in the trunk of the car," Fish recalled. "But finding rotting body parts in black plastic probably isn't anything new in New Jersey."

The flipper remained safely ensconced in a freezer in Pennsylvania until Fish found a student who was willing to work on it with him. They cut the flipper into one-inch steaks, took photos and analyzed the structure of the flipper (which was indeed covered with bumps known as tubercles on the leading edge).

Just as wings do through air, a good swimming appendage tries to maximize lift and reduce drag to get the most out of the applied thrust. One way a wing does that is to have a cambered wing—one with a curved surface on top, as Geoff Spedding explained earlier. Another is to increase its angle of attack and tilt the leading edge upward. This allows the animal or vehicle to make a banked turn. The higher the angle of attack, the sharper the turn. But again, there's a limit to how high an animal can tilt its fin before it stalls, leaving it dead in the water.

Most baleen whales, such as blue whales, are enormous animals and don't bother with the kinds of acrobatic maneuvers that might get them into this type of trouble. They tend to just swim straight and hoover up whatever unfortunate krill happen to be in their path. But humpback whales take a more active approach. These "aquabats" can swim in tight, circular corkscrews, creating a wall of bubbles to ensnare their prey. Then they turn and dive right into their bubble-net, scooping up mouthfuls of the trapped krill.

Humpback whales are large animals that can stretch to forty-two feet and weigh around forty tons, so this kind of agility is almost shocking. They are also the only species among the baleen whales that have these strange bumps on their fins; all the others' leading edges are smooth. And, as Fish soon learned, the tubercles may have a lot to do with their exceptional agility. These nobbles appear to be placed in a specific formation, and seem to get relatively more crowded the closer they get to the flipper tip.

The scientist found that the tubercles were actually helping to control the boundary layer in a surprising way. In the troughs between each tubercle, the flow was separating, which would theoretically be a bad thing. But as it separated, the straight flow turned into pairs of vortices, wheels of air swirling over the flipper's surface. As they turned, these vortices in the troughs actually helped keep the streams of water flowing over the tubercle peaks in line, even energizing them, in the way the wheels in a tennis-ball machine grip a tennis ball and launch it far and fast.

The end result? While the flow in the troughs turns into vortices that separate from the surface, those vortices hold the flow over the tubercles in place for longer, allowing the animals to execute maneuvers with much higher angles of attack.

Fish realized this shape, remarkably efficient for reducing drag and preventing stall, would actually be a great design for the blades of industrial scale fans and turbines, perhaps even vehicle propellers. They were less noisy and more efficient, which meant one of the companies that ended up using the technology could actually make fans with five blades instead of ten, saving both on material costs now and energy costs later. Fish actually created a company of his own, called WhalePower, with the plan of putting these blades into windmills. But it was the late-2000s, and the economy tanked; and the momentum for bringing the technology to market was largely lost.

Fish has been working on whales for about thirty years, and he still thinks that tubercled blades are a good idea with a host of potential applications, both in water and air. He even has a wind turbine working at the Wind Institute of Canada on Prince Edward Island, where he's found that the turbines really do perform better at moderate speeds compared to ones without tubercles.

"I think they have promise; are they the salvation to everything? No," he said. "But I think in certain niches, they will give us real benefits."

In the meantime, he's also worked on other forms of underwater motion, including helping to design a manta ray robot that could one day be a model for efficient, seafaring drones, whether monitoring ocean health or performing military surveillance. These aren't the only designs for swimming robots—scientists are buildings all kinds, based on fish or octopuses or even, as we'll discuss soon, robot jellyfish. None of them holds a perfect solution, but they each feature a particular advantage in a different environment. A propeller-based vehicle might be great for maneuvering down to survey underwater damage, such as oil spills. A more passive robot would be better for long-term environmental monitoring. Just as different animals have different niches and different advantages, so should the machines that are inspired by them.

WhalePower, Fish's company, takes as its motto: "Building the energy future on a million years of field tests." The point is, these innovations are inspired not by a particular ideal in nature, but by the diversity of its living things. And as climate change and human encroachment threaten these animals, we potentially lose our chance to learn from all of those field tests. Humpback whales are no exception: coveted for their meat and blubber, these gentle giants were nearly hunted to extinction in the early 1900s.

"If humpback whales had been killed off in the early 1900s, would there would have been a sculptor that had seen what they look like, and would I have seen those bumps?" Fish asked. "I might have only seen a skeleton in a museum. So it's up to us to preserve this, because we don't know what other plant, animal, or other living organism is actually going to inspire us or actually lead to some sort of new innovation. Who knows what the next technology will come from?"

"Jellyfish expert" is a strange career to land in when you're an aeronautics engineer who was born in Toledo, Ohio, a kid who expected to return to the Midwest after college. But John Dabiri, a Caltech engineer turned Stanford professor, is a surprising marriage of apparent contradictions. The mechanical engineer, winner of a 2010 MacArthur genius grant at the ripe old age of thirty, is a devout, church-going Christian who has to reconcile the Bible's teachings with Darwin's theory of evolution. He's also the child of Nigerian immigrants who fled political unrest, ending up in a

Midwest town where race lines ran along black and white, a world into which he did not quite fit.

At home, his world was governed by a very different set of rules. Dabiri's father was an engineer who taught math at a local college and his mother was a computer scientist who built her own IT company. To escape the pressures of their fraught neighborhood, they sent him to a small Baptist high school, where he graduated at the top of his class. Dabiri always thought he'd end up working for one of the great American car manufacturers, like General Motors or Ford Motor Company. It would be a good, straightforward, honorable way to make a living. But one of his professors at Princeton, Alexander Smits, pushed him beyond his comfort zone, encouraging him to work with Morteza Gharib at Caltech.

Gharib had already found himself looking in very different places for inspiration: he'd explored how embryo hearts pump fluid and the turbulence formed around combat aircraft wings. He encouraged Dabiri to also start thinking beyond car engines and jet turbines. Dabiri took to the idea like a fish takes to a bicycle. He hated it. He'd spent the last two summers while at Princeton just working on helicopters. To his mind, biology was little more than "memorization and stamp collecting," as he later put it.

Gharib was not deterred. He persuaded Dabiri to abandon a prestigious internship at Ford Motor Company and take the Caltech internship instead. (In some acronymic foreshadowing, the program was called a Summer Undergraduate Research Fellowship—or SURF for short.) Dabiri shrugged—at least, if nothing came of it, he'd have spent a summer in sunny California. He boarded a plane headed to Los Angeles. It was, perhaps ironically for a helicopter engineer, the first flight of his life.

The elder scientist took Dabiri to the Long Beach Aquarium

in search of inspiration. There, in front of a tank filled with jellyfish, the would-be rocket scientist found himself drawn to the animals' slow, pulsating movements. The jellyfish is a remarkable creature. It has a soft body with no armor and no brain—just a diffused central nervous system. It has no anus, so what it doesn't eat must come back out the same way. It lacks the sleek, powerful silhouette of the finned fishes whose speed, strength, and agility have allowed them to dominate the oceans since the Devonian age some 400 million years ago. And yet, the jellyfish has survived, even thrived, for 650 million years—well past the dinosaurs that rose and fell in the intervening eons. In some cases, the tentacled creatures have performed too well, spreading into warming habits and disrupting ecosystems. Some estimates say that some 40 percent of the biomass in the ocean is made up of jellyfish bodies. In any case, the jellyfish must be doing something right, the engineer reasoned.

"On the one hand they looked very simple, but there's a lot of interesting complexity there," he said later of the experience.

That day, Dabiri watched the jellyfish's bodies. But in his research, he shifted his attention from the animals' movements to their wakes—specifically, on the vortex trail they left in the fluid. The young engineer couldn't help but notice the potential similarities between the jellyfish and the human heart. Both were essentially just muscles to pump salty fluid. He and Gharib studied the vortex trail left by the jellyfish through the water and compared it to the trails left in blood flowing into the heart's ventricles. That research proved to be revelatory. They soon learned that vortex signature left in the blood flow could quickly reveal whether the heart was pumping normally or not. This could tell cardiologists whether their patients' hearts were healthy or damaged long before other symptoms start to show.

That summer changed the course of Dabiri's academic life. He returned to Princeton and applied to Caltech for his doctorate, with Gharib as his mentor. He began probing the jellyfish's secrets from all angles. He watched videos of jellyfish and tried to build physical models (essentially rudimentary robots) that would generate wakes that were similar to the animal's. Once he started thinking outside the engine, it was difficult to stop. Equations don't care about cosmetic differences between a jellyfish and a jet turbine, Dabiri realized. They describe patterns in the flow. Those patterns can be found in air or in water, and can be made by living things or by machines. The aerospace engineer began applying his jellyfish equations to submarines to make them more efficient—not necessarily by turning the metal vessels into rubbery sacs but by tweaking the design in order to modify the eddies and flows they created in their wakes to look more like the animals' vortex trails.

Dabiri started looking to other ocean-dwellers for inspiration. He noticed that schools of fish stay in these tightly packed groups—even though they shouldn't be able to. The turbulence generated by one swimming fish should seriously trip up the fish swimming behind it. In reality, each fish can swim more efficiently by staying in the right spot among the swirling vortices created by the fish ahead of it than it could by swimming alone. It's similar to the problem faced by wind farms—one wind turbine wrecks the airflow so badly that the next wind turbine often needs to be stationed a mile away. Dabiri took the tools biologists use to calculate ideal school positions and started using them to model diamond patterned turbine placements. By switching to vertical-axis turbines and positioning each one as if they were schooling fish, the engineer found that each turbine could take advantage of the turbulent flows created at just the right point to actually help their performance, rather than harm it. The discovery could improve the power-per-acre yield of wind farms tenfold.

There are some critiques of the vertical-axis wind farm. Mike Barnard, a senior fellow at the Energy and Policy Institute, has argued that the potential of vertical-axis turbines has already been explored more than twenty-five years ago by researchers associated with Sandia National Laboratories, among others. More to the point, he's written that Dabiri's work relies on a faulty assumption: that the real estate between the traditional horizontal-axis turbines on a wind farm can't be used for anything else. In fact, he says many turbines simply take up the small patch they're built on, and that small patch can be leased out while the rest of the surrounding area can be used for other purposes, such as agriculture. If you look at it that way and do the math accordingly, horizontal-axis wind turbines are far more efficient than vertical-axis wind turbines—which is why they've continued to dominate the market.

"I admire his conviction, but he reminds me of the prognosticators in the early twentieth century who insisted that automobiles were a frivolous idea and could never compete with the locomotive," Dabiri told me in an email. "His argument comes down to the fact that the current technology works, and new ideas like VAWTs haven't been proven yet. There is no underlying technical flaw that he can point to in the new ideas, only the lack of precedence for new wind technologies being successful."

Very little research has been done on how to arrange the eggbeater-shaped vertical-axis wind turbines in the last three decades, he says. And as far as he's concerned, leaving all that space between the pinwheel-like horizontal-axis wind turbines is basically just leaving energy-generating potential on the table.

In developed countries such as the United States, our energy is generally centralized—produced at a hydroelectric dam or a coal-fired power plant or yes, a wind farm—and then sent via wire over potentially hundreds of miles to homes and businesses. But in more rural or remote parts of the developing world, decentralized energy, where the power-producing plant is nearby, might make much more sense.

"So rather than having a huge plant out in Texas or the Dakotas that's generating hundreds of megawatts and then transmitting that over a thousand miles to the place where it's actually consumed, I think there's a lot of opportunity, a growing opportunity, to generate energy close to the end user," Dabiri says.

Vertical-axis wind farms would fit that bill much more neatly. They're shorter, on the order of thirty feet, while horizontal-axis windmills can reach ten times that height. That makes them less obtrusive, easier to place near the homes and businesses to which they provide power. It's an intriguing idea—but one for which evidence is currently lacking.

To that end, Dabiri has set up an experimental wind farm to test his theories—in Igiugig, Alaska, a windy tiny fishing village with a population of about seventy people. It's cold, windy, and remote, and so the residents have to fly their much-needed fuel in. Energy, per kilowatt-hour, costs them around four times the national average. The scientist hopes that vertical-axis wind turbines, properly configured and arrayed, could end up covering most of the village's energy needs, freeing them from this near-complete dependence on polluting hydrocarbon fuels.

Part of that research, done in partnership with Jifeng Peng of the University of Alaska Anchorage, involves testing turbines from different manufacturers, to see which ones perform best for his project. After all, he pointed out, hardly any research has been done on vertical-axis wind turbines since the mid-80s.

"We've been starting almost from scratch in some cases,

trying to reinvent some of the design features, figure out what didn't work thirty years ago and how to make it work now," Dabiri says.

Working in such a remote area has its challenges. They have to figure out ways to keep the turbine blades from getting iced up, which can affect their performance. They've only put up three turbines from two different manufacturers to see which one works better—because if there are problems it's better to just repair one or two rather than ten at a time, especially when supplies need to be brought in from as far as China. But they're expanding it to five or six by the end of 2016 and then planning to put up a few dozen the following year.

There's another research project in Igiugig run by a group at the University of Washington, which uses what Dabiri describes essentially as a vertical-axis wind turbine on its side to generate hydrokinetic electricity. This could potentially provide 30 percent of the village's required energy, while the schooling turbines, in its most ambitious configuration, might cover 60 to 75 percent of their energy needs.

"In a perfect world, my vision would be to convert this village to an all-renewable system," Dabiri says—one free of diesel fuel.

On a whim, I ask Dabiri if he's ever heard of Frank Fish's whale-tubercle work—and whether those could potentially be integrated into his schooling-fish turbines.

"Potentially!" he says. "In fact I think it probably makes more sense on a vertical-axis turbine than on a horizontal-axis turbine, frankly." That's because the blades on a traditional windmill cut through the air at fairly low angles of attack—which means they're not taking full advantage of the tubercles' ability to help humpback whales maneuver those sharp turns. Vertical-axis turbines, on the other hand, do end up cutting through the air at high angles of attack at certain points in their rotation, and so placing those nobbles on the blades' leading edges could raise their efficiency even further.

From a variety of sea creatures, Dabiri has been able to draw out insights that could impact fields as broad as medicine and climate change. And if his life seems charmed—tenured professor by twenty-nine, MacArthur genius by thirty—he never seems to take his success for granted. Facing failure and rejection is part of the gig, he says. He's had significant NSF grant applications turned down even as he's been lauded in the prestigious *Nature* magazine, and papers rejected by major journals even as he's won career-defining awards. When the MacArthur Foundation first contacted him, he assumed they were just looking for Gharib.

Certainly, some of that rejection has changed to respect over time: When he first proposed the jellyfish-submarine idea, a naval officer laughed at it. Now he's getting funding from their research office. As for where that half-million-dollar genius grant has been going, the scientist said he'd set a little bit aside for some more humbling jellyfish-inspired research: swimming lessons. It's true—though he has an aquarium in his lab and studies animals from the ocean, Dabiri has never learned how to swim.

I check in with Dabiri with some regularity, as I often cover his research, and recently asked him how those swimming lessons were going.

"Well, I can sink calmly; that's about what I've gotten to unfortunately," he says, with a sheepish laugh. "So at this point I've decided I'll just spend my time trying to understand how the good swimmers accomplish their task in nature . . . and maybe eventually some of these ideas would rub off."

Perhaps the most riveting presentation I saw at that 2010 fluid dynamics conference—and there were many—was a presentation on flying snakes.

Yes, you read that right—flying snakes. Or perhaps more accurate, gliding snakes: serpents that can, from a treetop or other platform, glide as far out at 330 feet. It sounds like the stuff of nightmares, worry not: these snakes are found exclusively in Southeast Asia, and you are unlikely to encounter them unless you go looking for them. Nonetheless, these animals seem to bend the very rules of nature. Most aerodynamic non-avians—flying squirrels, gliding lizards—have large winglike flaps of skin that stretch from arm to leg, creating a kite-like structure they can use to sail to safety. But snakes have no such limbs from which to hang such a handy chute. Perhaps the freakiest thing about these snakes (who belong to the genus Chrysopelea) is that they look deceptively ordinary.

However, if you build a platform in the middle of a forest clearing and then film a snake leaping into the air, you'll see there's nothing ordinary about the way they move. At first, after launching themselves from the platform, the snakes plunge into a steep dive—but then they just as abruptly level off and continue on in a shallow glide. While airborne, their body goes through a number of brain-bending poses, in which the snake's back half almost seems to overtake the front half, sending it nearly tumbling. But it never does; the tail does some whippy things and then the undulations start to make a kind of strange sense. As Samuel L. Jackson might have said in a slightly different movie, these mother-effing snakes don't need no mother-effing plane.

John Socha, a biomechanist at Virginia Tech, has made a career out of studying these mysterious animals and decoding the many adaptations that go into their flying ability. First off, the animals actually leap into the air: They wrap the backs of their bodies around a tree trunk or a branch and then launch the fronts of their bodies forward, making them the only snakes to execute a true jump. Then they flatten their bodies, doubling the surface area of the bottom, and giving the top an aerodynamic, parabolic curve. They curve their bodies in a way that turns the entire surface into a sort of wing; but, because it's an asymmetric wing, they keep undulating side to side, to keep their bodies from tipping to one side or the other.

Socha, in the middle of his presentation, then turned to a video that was as horrifying as it was hilarious. He and his colleagues had designed the arena so that the snake would want to fly straight, toward apparent escape. The researchers would stand safely off to the side and run out to grab the escaping animal once it had landed. But in this clip, the snake launches itself and then immediately hooks a sharp left—straight toward the two scientists waiting in the wings. One dives behind the scaffolding while the second blue-gloved assistant looks up in disbelief before ducking out of the way.

I have to admit, every time I see this clip it makes me chuckle, thus reaffirming that I am a terrible person. But I also rewatch it for another reason: Because it blows my mind that these snakes can actually steer. These animals "fly" using a completely different method from those we're used to seeing in birds, bats, and gliders, and analyzing and modeling how they manage it could give us unimagined insights into how life navigates a fluid-filled world.

PART III

ARCHITECTURE OF SYSTEMS

5

BUILDING LIKE A TERMITE

What These Insects Can Teach Us About Architecture (and Other Things)

The Eastgate Center in Harare, Zimbabwe, lies on the figuratively problematic intersection of Robert Mugabe Road and Sam Nujoma Street. From a distance, the grayish structure looks like any other concrete mixed-use office block you might find in southern Africa, its wide, squat frame contrasting with the occasional bluish skyscraper punctuating the skyline. But as we get closer, the details begin to set it apart. The brushed concrete has been made to look something like granite, in an homage to Great Zimbabwe, a ruined city built of stone between the eleventh and the fifteenth centuries by the ancestors of the Shona people. The slabs of concrete that make up the building's frame are actually cut into X-like designs, giving the eight-story structure covering half a city block a surprisingly airy feel, as if it were a trellis or an arbor, waiting for some vines to grow. And there are indeed plants climbing the walls inside and out—a natural, built-in cooling mechanism. Pigeons also make their nests in the many nooks and crannies around the building.

"I like the birds," Mick Pearce tells me as we enter the building. "There's a peregrine eagle on the other side."

Eastgate stands as a testament to the benefits of architecture inspired by nature. Completed in 1996, it was designed by Pearce, an architect who was fascinated by the termite mounds surrounding his home, whose tiny inhabitants seemed to regulate the internal temperature without needing the costly heating and air-conditioning equipment that human buildings seem to require.

The prototypical office buildings that define city skylines are the skyscrapers, mirrored surfaces reflecting their surroundings like a Hollywood starlet's aviator sunglasses. To me, they represent what was thought to be beautiful in the 1990s—a *Matrix*-like dreamscape of dead metal and glass, divorced from the nature environment, the perfect setting for an unreal cityscape created by robots to control humanity. Their beauty lies not in an expression of architectural skill but the flat rejection of biological limits.

Whether tall or short, these office buildings are fundamentally energy inefficient. In the winter, they bleed out their warmth through those wide glass windows, requiring heaters to make up the difference. In the summer, they act like greenhouses, baking the residents unless they run the AC all day. These buildings are constantly fighting the natural climate, working against it instead of with it. They also recycle already-cooled air to save energy costs, which could potentially contribute to indoor air quality problems.

These buildings, almost willfully inefficient in spite of triple-glazed glass and other technological improvements, have major consequences for the environment and the economy. Buildings account for 40 percent of the United States's total energy consumption—about 21.4 percent residential and 18.6 percent business. In cities like New York and Chicago, where an increasing share of the human population will end up in the coming decades, that share shoots up to around 70 percent. Much of this is avoidable and inefficient use of heating and air-conditioning;

some of it is exacerbated by human behavior. For example, a recent survey by the Natural Resources Defense Council of three hundred retail businesses in Manhattan and Brooklyn found that 20 percent of them would run the air conditioner and leave their doors open on hot summer days to lure in customers, leaking loads of "coolth," raising their monthly energy bills by up to 25 percent and dangerously overloading already-strained municipal power grids.

As concerns about inefficient buildings have risen, the tide has begun to turn against the glass edifice. Ken Shuttleworth, one of the architects behind the iconic "gherkin" building in London—which looks rather like a giant pointy Fabergé egg—has rejected the skyscraper. "We can't have that anymore. We can't have those all-glass buildings," he told British media in 2014. "We need to be much more responsible." Just months before, another oddly shaped London skyscraper nicknamed the "walkie-talkie" had been caught frying nearby storefronts and melting the side-view mirror off of a parked Jaguar; the curved, reflective edifice focused the sunlight onto the street below, burning carpets and causing plastic to bubble.

It's become clear that if cities are to meet stringent U.S. and international goals for cutting greenhouse gas emissions in the next few decades, we have to make the city's basic units, its buildings, far more energy efficient than they are today. That will take some pretty radical rethinking in architecture, engineering, and construction.

Eastgate was a trailblazer in this regard, an architectural retort to the glass skyscraper's design problems. The mall lacks an air-conditioning system, and yet manages to stay relatively cool in the summer and warm in the winter. It does so in part by taking advantage of what's known in building engineering as the stack effect, moving warm air up through and out of a column by dint of its buoyancy, thanks to differences in temperature and moisture between the air in the column and the air outside. This, for a long time, is what scientists believed was behind the shape of certain termite mounds with "open chimneys"—the hollow in the middle, they assumed, was meant to keep the nest, hidden below the mound, comfortably cool.

Pearce's design for the concrete office block used fans to pull cold air in near the bottom of the building, run it through the rooms to cool them, and let it escape through the chimneys in a central ventilation system. The architect estimates that the building saved about 90 percent of the energy costs compared to a conventional building. The design has been shortlisted for and won a number of international awards, including the 2003 Prince Claus Award for Culture and Development.

There's a slight catch though, a biologist named Scott Turner tells me hesitantly, as if he's afraid of breaking bad news. While Eastgate's stack-effect design was supposed to mimic termitic architecture, that's not how termite mounds actually work.

Three hours north-northwest of Windhoek, Namibia's capital city, many miles past triangular highway signs warning of warthog crossings and hillsides blanketed in colorful tin-roofed huts, Scott Turner's boots crunch through the dry grass and cruel acacia thorns in the fields behind a ranch house owned by the Cheetah Conservation Fund. It's quiet out here in a way I rarely experience—the wide silence that comes from being so far outside of a city that all the noises you never notice, like the constant sticky susurration of rubber tires rolling over road, are suddenly, unmistakably absent. In their place: the occasional helicopterlike whirring of giant, jewel-like dung beetles, and the raucous cocktail-party chatter of unseen, unfamiliar birds jabbering at and over one another.

Turner, a physiologist at the State University of New York, heads toward a brick-red cone rising out of the earth like a wizard's hat. It's tall; he only has to stoop slightly as he leans in to examine it, pressing his palm against the sun-warmed slope like a doctor checking a patient.

Every year, Turner and a group of researchers of all stripes head to southern Africa, to visit a country that holds a rich diversity of mound-building termites. Among his companions on this trip: two engineers, two physicists, and an entomologist with an interest in bioinspired robotics. Previous trips included other researchers who studied the termites to program robots who could build structures without requiring a central plan or guiding architect. Like termites, these robots just followed a few simple rules that, as they interacted with each other, allowed them to build a structure within certain design specifications. While I was covering the robotics work for the *Los Angeles Times,* one of the Harvard engineers mentioned working with a SUNY biologist—which is how I found out about this trip in spring 2014, and begged Turner to let me tag along.

Scott Turner came into termite research by accident. A physiologist by training, he had always been interested in the interface between organisms and their environment—a border that he has increasingly found to be far blurrier than one might assume. Turner spent a few years in South Africa in the late 1980s, spending one year teaching at a very remote campus up near the Botswana border. Termite mounds dotted the grassy landscape, and the professor decided to use these abundant structures as a classroom demonstration, to show to his students some principles of measuring airflow. (This was partly out of desperation, as someone had apparently made off with all of the biology lab's equipment, save some flasks and a water bath.) He pumped a little propane gas into the mound and put a sensor in the mound's chimney, to see when the gas made its exit. Using this system, he'd expected to see air flowing steadily through large tunnels in the mound—and was perplexed when the probes revealed no such steady streams of air passing through the conduits.

"I was relying on what everyone said in the literature on how they had to work . . . and the airflows weren't anything like what everyone said they have to be," Turner said. "So that got me kind of intrigued."

The first thing you might want to know about those majestic termite mounds is that termites do not live in them. Instead, the insects live in an underground nest, where they grow gardens of a specific fungus that help compost the tough wood that they bring home to eat. Termite mounds also come in a variety of shapes and sizes, including (but not limited to): gently rounded mounds; tall mounds with pointy spires; conical mounds with an open chimney at the top; tall, skinny columns; and shorter cylindrical "chimney"-shaped mounds, also open at the top. Some mounds in India almost look like fairy-castle fortresses, with a spire in the middle and buttressed peaks surrounding it. I imagine this diverse array of termite-sized skyscrapers in one spot—they'd make one heck of a skyline.

Surely such complex structures must be built with a purpose in mind, scientists thought. Back in the late 1950s, Swiss entomologist Martin Luscher presented his idea for the function of termite mounds in *Scientific American*: They were being used to help regulate the termite nest's collective body temperature. After all, a termite nest's metabolic rate runs pretty high, somewhere around 55 to 200 watts—about as much as a goat or a cow. If the termites didn't somehow manage to vent some of that heat, they'd get baked pretty quickly. But termites can't let all of their heat (and moisture) out into the environment—they're delicate creatures, unable to survive out in the open. And so the mound, the thinking went, was a way to regulate temperature (and humidity).

Termite mounds come in a wide variety of shapes and sizes; Luscher said he examined those of *Macrotermes natalensis,* which made a rounded, closed peak of a mound with ridges running down the sides, giving it a craggy exterior. (Turner has suggested that Luscher might have been looking at *Macrotermes bellicosus,* which builds similar mounds.) This edifice doesn't have any large openings through which to vent heat—but the mound is shot through with large tunnels, called surface conduits, that run parallel just beneath the structure's surface. Out of the nest, which lies beneath the ground, rises a thick "chimney" of sorts, which connects to a complex "reticulum" of tunnels that link up to the surface conduits. The surface conduits also connect to the egress complex, a tangled nest of tiny tunnels that open onto the surface, making it porous.

According to Luscher's model, the air near the nest was heated and humidified by the termites' activity, thus becoming more buoyant and then rising through the central "chimney" connecting the nest to the mound. But the top of the mound is closed—the air has nowhere to go but continue through the tunnels that run right beneath and parallel to the mound's surface. As it passes through these tunnels, within a few centimeters of the mound's sloped surface, the air is "refreshed" though the porous walls by air filtering in from the outside. Now cooler and drier, the air gets denser, and continues sinking down the conduits, which wrap around and lead back to the nest, where it's breathed in and again heated and humidified by the termites. This model, then, describes the mound as something like a heart-lung machine—it outsources the termites' oxygen circulation. This model is known as thermosiphon flow.

But what about those mounds that had an open chimney at the top? A different theory arose to explain the dynamics in these Macrotermes mounds, called "induced flow." In the model, wind blowing over the top of the mound pulls air out of the top of the chimney, helping to pull the air under it upward, too. Vents around the bottom of the mound pull fresh air in at the bottom, through to the nest, where it picks up heat and then rises up through the chimney, and pulled out with the help of steady winds. To architects, it's better known as the "stack effect," and it's clear that this is what inspired buildings like the Eastgate mall, which features a series of chimneys that help vent warm air that builds up inside the structure.

In the decades since these theories arose—and especially since the building of the Eastgate shopping center—many experts have extolled the temperature-controlling virtues of termite mounds, claiming that the mounds are always kept within one degree of their ideal temperature.

"It must be kept at exactly 87 degrees, while the temperatures on the African veld outside range from 35 degrees at night to 104 degrees during the day," a 1997 *New York Times* story on Eastgate explained.

But, as Turner began to realize after his ill-fated demonstration with his South African students, this wasn't really true. Tracking mounds from early 2004 to early 2005, he found that the *Macrotermes michaelseni* mounds varied in temperature throughout the year from roughly 14 degrees Celsius in the winter to 31 degrees in the summer—a variation that tracked with the temperature of the soil they lived in. That fundamental assumption—that mounds' airways were designed to maintain a constant temperature (and humidity), acting as a natural "air conditioner"—simply could not be true.

From 1995 to 1997, he studied forty-five closed mounds built by a species called *M. michaelseni*, which look somewhat like the structures created by *M. natalensis* but without the dramatic ridges. He injected them with a propane gas tracer and monitored how it moved through the tunnels by inserting sensors all over the mound. He soon showed that the air wasn't circulating through the mound the way that Luscher had described, moving in this orderly circuit powered by the nest's heat and damp. So much for the heart-lung machine.

Instead, what was happening inside the mound was not circulation, but mixing—a messy, tumultuous encounter between two bodies of air, the old stuff hovering above the nest and the new stuff somehow making its way in through the mound.

Turner realized that the air's movement had little to do with thermal regulation. In reality, the mound was actually acting more like a real lung—facilitating the exchange of oxygen and carbon dioxide, so that the insects wouldn't suffocate in their own exhalation.

Turner has done a number of experiments, including injecting propane into the tunnels in the mound and watching what happens to that tracer gas. He's found another explanation for what's powering this mixing in the mound—one that isn't powered mainly by buoyant air (as in the closed-mound model), or by steady wind aided by buoyant air (as in the open-mound model). Unlike these models, his explanation doesn't involve an orderly, conveyor-beltlike flow of air through the mound's major passageways.

Instead, Turner concluded that turbulent airflows—winds and breezes that accelerate and decelerate and change direction—were powering the mixing effect he was picking up within the mound.

For an engineer, this might seem like a strange concept. After all, turbulent winds are noisy and inefficient, and really just tend to get in the way. Think of a toy pinwheel: It's just as likely to get deformed by turbulent wind as it is to be propelled by it, which means that it doesn't do much except flail this way and that. To get a good spin going, you need a steady wind—which the pinwheel-holder often has to create by blowing into its wings.

Here's the problem, though: Just as the natural world doesn't have well-paved roads, it doesn't have predictable, steady winds—especially not in Namibia. If you built a structure to function with the use of steady winds, it wouldn't work; transient winds, somewhat ironically, are more easily relied on. ("The only constant is change," the saying goes.) The termite mounds seem to have built a structure that manages to harness those ephemeral winds in order to power the mixing that exchanges fresh oxygen with the old carbon dioxide.

The more Turner thought about it, the more it seemed to parallel the function of human lungs. Air passes through the trachea, which splits into the bronchi, which branch into smaller passageways called bronchioles. Lungs don't work by circulation; while they pull air into the main passageways (the trachea and bronchi), that's basically where the bulk flows of air stop. Bulk airflow may dominate at the start of the system, and diffusion at the end of it, but it's this middle zone, dominated by that mixing process, that is key. That's basically what Turner believes is happening around the base of termite mounds: the air in the nest is very still, not flowing; but trading carbon dioxide molecules for fresh oxygen ones across the border region that lies between the nest and the passageways closer to the mound's surface.

In the case of humans, we can pull air into the lungs to get that process going. Termites don't quite have that luxury—but

Turner doesn't think that the insects are passive actors in this process. Termites are constantly repairing and building up the mound, and he thinks they alter the shape of the mound, both inside and out, to take advantage of transient winds and power that mixing action.

Termites, like ants, live such cooperative, communal lives that they have been collectively described as "superorganisms." Each individual may seem dumb, and blind, and have no long-term goal or strategy of its own, but when they act together, they seem to make swarm-level decisions for the entire colony so seamlessly that each insect appears to be merely part of a limb in the body of a larger creature. This hive mind is a concept that fascinates humans, in part because this "mind" has no physical location, as it does for us, in our brains. And that ability—to take a simple set of instructions and make decisions that seem actually intelligent on a larger scale—was the focus of the Harvard engineers who built the termite robots, and is what draws other members of this spring's expedition. But we'll get back to that idea later.

If the termite mound really does act like a lung, then it's basically acting as an external organ for the superorganism—part of what's known as the "extended phenotype," a concept described in a book by evolutionary biologist Richard Dawkins in his 1982 book of the same name. Turner has spent his career studying the border between one organism and another, and one organism and the larger environment, ultimately discovering that these distinctions are not quite so clear-cut. The boundaries of a creature's "living" parts, in this sense, extend far beyond the limits of its skin (or shell, or exoskeleton).

Just as he blurs the line between organism and environment, Turner has also been chipping away at the boundary between biology and engineering. A few years after embarking on his termite studies, Turner teamed up with Rupert Soar, a researcher currently focusing on digital construction technologies and biometric fabrication at Nottingham Trent University. The pair wrote a paper discussing the erroneous scientific theory underlying the Eastgate model and described their own ideas about the mound's wind-assisted ability to "breathe."

"This is not intended to be a criticism, of course: Pearce was only following the prevailing ideas of the day, and the end result was a successful building anyway," they wrote, back in 2008. "But termite mounds turn out to be much more interesting in their function than had previously been imagined. We believe this betokens expansive possibilities for new 'termite-inspired' building designs that go beyond Pearce's original vision: buildings that are not simply inspired by life—biomimetic buildings—but that are, in a sense, as alive as their inhabitants and the living nature in which they are embedded."

In some ways, the fundamental problem the researchers have with Eastgate is that the design did not go far enough to emulate biology. There's a big difference between strapping two wings to your arms and flapping, and understanding every individual part of a bird and its overall form and the fluid dynamics around its body when flying. Only then can you apply those principles to modern airplanes, which are built at a different scale, out of different materials, to fly at far greater speeds. After all, you wouldn't try to analyze a poem in French and translate it into English, unless you fully understood the language. How could you expect to effectively translate the lessons of biology into practical man-made designs, without the same depth of knowledge?

The scientists have set out to do just that: performing much-needed basic research on several aspects of the termite and the mound, to get a full-resolution picture of the termite and how it builds. As the engineer to Turner's biologist, Soar sees their collaboration work on multiple levels. Turner looks at the mound as an extension of termite physiology, observing its structure from the bottom up, trying to understand how termites interact with each other and the environment to build the mounds; meanwhile, Soar takes the top-down approach, filling the mounds with gypsum to try to understand the large-scale structure and translate the findings into the framework of structural engineering. In the years since they've jumped into termite research, the boundaries between their roles and their interests have started to blur.

Soar is tromping through the grasses with Turner, a stethoscope dangling from one hand and a pipe clenched between his teeth, part of the diverse crew of researchers that has gradually accumulated around Turner over the years—part of his scientific extended phenotype, if you will.

"They're incredible earth-engineers," Soar comments, looking at the mound. In recent decades, researchers have begun to realize the impact of this tiny, terraforming insect. "They reckon the entire middle landscape of the earth is shaped by termites."

"Middle Earth?" jokes Paul Bardunias, an entomologist from the University of Florida doing his postdoc with Turner.

"Yeah, Middle Earth," Soar says, good-naturedly fielding the Lord of the Rings reference. "You start digging down, you find all the rocks and boulders about a meter underground, so all the soil they've poured over that over thousands and millions of years, the rocks all sink. And that's why you get the red soil."

"So it's not the blood that everyone's spilled," comments German engineer Max Kustermann, apparently in reference to Namibia's violent colonial history.

No blood, but perhaps other bodily fluids. Termites take bits of rock and debris, coat it in their feces and gooey digestive juices, turning it into an incredibly fertile layer of soil. They bring water up with the soil they dig out, making it available for roots that can't reach quite so deep, which is part of why an arid country like Namibia can host seasonal wetlands. Termites are quite handy for miners, too; the insects' excavations bring up deeply buried minerals, so prospectors can examine an area's termite mounds to see what lies beneath the earth. That's a much cheaper, quicker, and less risky proposition for companies than bringing in the equipment and personnel to dig around.

Each of the researchers at this ranch house at Cheetah View, some twenty-five minutes west of Otjiwarongo, the nearest city, is here for a different reason. Rupert and Max, a well-dressed engineer, have teamed up to look for innovative ways to bring 3D printing and digital fabrication tools to the process of building and construction. Bardunias studies how individual termites make decisions in an attempt to understand their programming—why they build how they do. His work could further inform the creation of more capable building robots. But Turner's long experience with these incredible builders, and his studies of their superorganism-like properties, acts as the throughline. On this morning, he's accompanying Harvard physicists Hunter King and Sam Ocko as they test their airflow monitors in the mounds.

Sam, a graduate student, sits on a folding stool he brought with him and opens up a laptop, wired up to a little box that looks like a modem, which itself is attached to another wire that ends in a little tampon-sized probe. Hunter, a postdoctoral researcher, is holding that probe in his hands while standing at the foot of the mound, perhaps seven feet away from Sam. Hunter tap-tap-taps against the mound, listening for a hollow spot that could give away the location of one of the surface conduits.

Termites do not live in the mound. They live beneath it, in a nest shaped like a flattened basketball, about 1.5 to 2 meters wide. That nest is founded by a single pair of mated termites that get busy making babies and, just four or five years later, have about a couple million termites in the colony with them.

Obviously, for such a quickly growing family, you're going to need some extra room. So the workers excavate the soil and dump it aboveground. They're not just trying to get rid of earth; they're also trying to bail out water, which is why they bring wet soil up to the surface, depositing it in a pile that eventually is sculpted into those impressive mounds.

"It's a dynamic structure; termites are always cycling soil up, mostly in the wet season, because the main impetus for building the mound in the first place for moving the soil upward, is water that percolates down," Turner said. "They're essentially getting rid of excess water in the nest."

Once it's up there, they keep adding to it by bringing more wet soil to the surface—which means that they need a whole network of branching tunnels that allow them to move the soil all the way to the outside. So the pile is a kind of rough draft that is eventually refined, with pathways carved into it over time. Certain roads are widened until they become highways, running right beneath the surface, called surface conduits. A tangled network of tiny tunnels, known as the egress complex, breaches the remaining half-inch to an inch between the conduit and the mound's surface. These little holes give the mound the porosity that allows it to interact with wind in surprising ways, turning it into a kind of respiratory structure.

The surface conduits actually extend underground, surrounding the nest, and then merge into an extensive network of tunnels that can extend up to 230 feet.

"Wait—there're tunnels out here?" I say, looking around my feet, standing some fifteen to twenty feet away from the mound itself.

"You're standing on top of a network of tunnels that are

foraging tunnels, and that's how the termites get out to find their food," he says. He points to a nearby tree, with red soil that seems to climb up onto its surface. Termites can't survive in the outside environment, he explained; it would be like expecting the cells in your body to survive without your skin. So wherever the termites go, they bring their soil-skin with them.

"They almost never are exposing themselves," Turner says. Once they build up the sheeting on a tree, for example, they'll mine the bark on the trees and then haul it back to the nest, where they digest it, defecate it, and then use it build a structure called a fungus comb, which they inoculate using spores from the environment. The spores germinate and the fungus spreads through the wood-turd structure the termites build. Wood is tough: Fibers of the sugar cellulose are held together with a gluey protein called lignin. These fungal combs help separate the fibers, breaking the complex sugars down into a product called hemicellulose, which the termites can easily consume. (This is different from other animals like leafcutter ants, which actually eat the fungus itself rather than gobbling the sugars it produces.) We'll actually get to see these combs tomorrow, he adds, when they break open a live termite mound.

For the moment, Hunter's having some trouble getting his probe into a surface conduit, one of the large main tunnels running underneath the surface. The physicists are looking to measure the large-scale airflows in a conduit, but they keep ending up in the thicket of egress complex.

"The problem we've been having is that sometimes when it sounds hollow, it's just a few of these skinny ones," Hunter says.

Turner bends over and raps his knuckles against the mound, moving across the surface. "Follow the hollow," he tells Hunter. He knocks, listening and moving upward along an invisible path. "Could be here, hold on a second."

Hunter inserts the probe. It works. "Phew!" Turner says, half-joking.

"Vindicated!" Soar says. "That was a master class, that."

Hunter and Sam are working under Harvard applied mathematician Lakshminarayan Mahadevan, who studies the physics of natural systems. The hope is that the physicists should be able to properly quantify the physiological processes that Scott Turner and Rupert Soar have been describing all these years. The physicists, in turn, get an interesting natural system to model. Turner and Soar are generally staying out of their way, in order to keep from biasing the researchers' results with their long-held ideas of what's going on inside the mound. This, of course, means that they must be prepared if the physicists' data shows that Turner's theories on the mound's relationship to wind—and perhaps even their function—which have held for all these years, may not be exactly right.

Earlier that morning, before we tromped out to the nearby mounds, I'd munched on my breakfast of fruit-and-nut muesli while watching Rupert and Paul take long coils of aluminum wiring and cut them down to short snippets and dump them in a bucket.

"This is where bio meets engineering, I suppose, in the world," Rupert says through the pipe clenched between his teeth, as he cuts wire down to size.

I soon realized that the line between biology and engineering can become apparent in surprising ways—simply in the words that are used. Rupert mentioned a few other researchers who were using 3D printing to explore the "materiality" of biomimetic materials.

"That's part of the fun of this interaction we have," Paul jumped in, "because you guys have these words that are just, like, magicky words."

"I know, I know, but so do you! I've had to learn all of yours," Rupert says.

Now, after examining the mounds with Hunter and Sam, it's time to see where all that aluminum went. After testing the mounds, the scientists and engineers tromp back over to the yard in front of the ranch house, where Stuart Summerfield, a technician who works with one of the biologists, has been heating a container to several hundred degrees Celsius—hot enough to melt those metal wires. Near Stuart's feet, embedded in the ground, sits a strange circle with a pattern of wavy holes in it resembling the ink splotches in a Rorschach test. As it turns out, the scientists had actually sawed the top off of a nearby termite mound, dried it out for several days, turned it upside down and buried it in the earth.

Stuart, his face shielded by a welding mask, uses thick-gloved hands to pour the silvery liquid into the biggest of the mound tip's Rorschach-like holes. There's a wet muffled boom—a sign that the mound top was perhaps not as dry as thought—and he quickly steps back, but keeps pouring, until the molten aluminum is spent.

Now that the danger of explosion is over, Turner and the other scientists cluster in, looking at the glaze of aluminum that's dripped onto the edges of the holes. There's still plenty of room left—looks like they might have needed more metal. But even if they manage to heat enough up to the required temperature of about 1,200 degrees Fahrenheit, it's unclear if it will be able to get into all of the nooks and crannies still left, given that the first batch of aluminum will have already solidified inside, blocking off access.

While the researchers discuss their options, Paul and Lisa Margonelli, a writer who's working on a book on termites, watch the interactions between anthills and termite mounds a few feet away.

"Where there's termites, there's ants," Rupert had said earlier this morning. But the two species are not friends. Far from it. Ants conduct raids on termite mounds, and if a termite gets a chance, it will drag a scouting ant into a hole in the termite mound; like a frat guy in a horror movie, that unwary ant will never be seen again. It makes sense—the two species are foragers looking for largely the same resources in the same environment. What may surprise you is that the two species, similar as they may seem in their appearance and communal behavior, are not closely related; in fact, the termite's closest living relative is the cockroach. In some ways, that makes it even more remarkable that they both seem to have a sort of hive mind, which is why scientists interested in artificial intelligence are studying both species. But I'll say more on ants, and artificial intelligence, in another chapter.

A harvester termite that's probably half the length of my fingernail has grabbed a twig that's probably eight to ten times its length, and is trying to lever the thing into a skinny hole in the ground. Paul and Lisa watch intently. If you've ever tried to maneuver a long sofa through a narrow, poorly positioned doorway, you've had a sense of this termite's struggle.

"Take it down, take it down," Lisa says, almost under her breath.

Our laser focus pulls Scott and Rupert away from the aluminum.

"He's measuring the length, look," Rupert says.

"He's measuring the center of mass is what he's doing," Paul said.

"You're kidding me," Rupert says.

"That is a crazy termite," Paul says. "Look at that . . . looks like you're right, he's chopping it."

Berry Pinshow, a comparative physiologist and the expedition's self-appointed parade-rainer, takes one look and dismisses the insect's hard work. "If he was planning to pull it into the burrow . . . stupid place to chop it off," he says, before walking away.

But sure enough, the termite flicks the stick around and pulls his prize into the hole.

"Guess what? Termite's smarter than Berry," Paul jokes.

Paul is looking at the behavior of individual termites, trying to break down their simple rules of behavior. Earlier that morning, while they'd been chopping up the aluminum, I'd asked him what his focus was.

"I was brought on because my thesis was on how individual termites come together—the algorithm they use to come together to create tunnels," he says. "What I'm doing is looking at individuals in Macrotermes and what are the rules they follow as a group, the interaction rules to create these mounds. And my job is to take them back to Harvard and tell them how to make robots that follow those rules."

I'm a little confused at this point. "So you're a biologist—an entomologist. So at what point does an entomologist think he's going to start dealing in a lot of algorithms?"

"These days, it's all about algorithms," Paul says.

"The world is dealing with algorithms," Rupert adds.

Paul is quick to point out that when he says algorithms, he personally means it in a rather cruder sense.

"I am not a computer programmer in any way," he says. "But I watch termites, and termites essentially are little computers, so they have a program."

If you can tease apart the behaviors in that program, then you know the code—which you can then program into computer programs, or even robots, that build those structures. If you understand how each termite works, and what happens when they interact, then you understand the entire mound.

"You're looking for what is a kind of fundamental algorithm," Rupert says. "Which may not exist but you're drawn to it because you're human."

"A holy grail?" I ask against my better senses, because I'm a journalist and we're irresistibly drawn to anything that can be called a holy grail. But the question seems to hit Rupert differently. I was thinking of it as the key that could solve many of the issues they've had trying to understand these creatures. But the engineer seemed to take a different, and perhaps more accurate meaning—a quest for an object that is never really found.

"It's probably a holy grail," he says after a pause. He laughs and turns. "Isn't it Paul? But we're deluded enough to believe that it isn't."

Now, as the aluminum cools and the group splits off to do their various tasks, I follow Paul through the yard, past the fence to the second of the ranch's buildings, where half of the researchers have their sleeping quarters. In the yard outside of these quarters, near the shade of an acacia tree, Paul has set up one of his experiments. Near a termite mound stand two Plexiglas plates roughly the size of printer paper. Between the plates, which are sandwiched together with binder clips, round blobs of soil appear to be slowly growing upward. Paul has tapped into a tunnel from the nest that leads into the narrow layer of space between these two transparent plates, giving the insects an almost flat space in which to build. This helps in two ways: one, it allows him to see the termites at work, rather like an ant farm would; and two, it simplifies their behavior, since they can only build in two dimensions—thus allowing Paul to try and isolate some of the building rules going into their design.

Paul had set up the plates around 11 p.m., so the termites must have built all through the night. This morning, they've managed to build what looks like a cartoon version of a chubby foot, with a large, lopsided, slightly pointy dome forming the body and bubbles growing from the top like fat little toes. The second toe, in fact, has managed to make it all the way to the top of the Plexiglas sandwich, with the big toe about an inch lower.

There are cues from the environment that Rupert and Paul think are noteworthy—moisture being one of them. Paul has put termites on soil placed between a dry wick and a wet one—so that one end of the ground is parched and the other end is soggy. The termites seem to start building in a goldilocks zone that has just enough, but not too much, water locked in. A certain moisture gradient, then, might be one of the insects' driving parameters.

Wind turbulence might be another one. Paul points to the "second-toe" of reddish dirt that has reached the open top of the Plexiglas sandwich. That contact actually seems to bisect the width of the transparent sheets—and Paul suspects that's no accident, that the termites sensed the wind patterns and were directly responding to cut the wind turbulence down.

"They may have broken the cycle of turbulence just by putting these flying columns up," he explained. "The placement in the columns could somehow correlate to eddies in the turbulence."

Paul wants to get the mound's growth curves by measuring the build happening between these plates over time. There are stages to the termites' construction process: A termite decides it's time to build somewhere, and puts down a mouthful of soil. Most believe it also lays down some chemical pheromones that will tell any passersby, "Hey! Put your soil down here!" This recruits more termites to dump their mouthfuls of dirt down in the same area, creating a quickly growing "bubble" of material. Later, after this "scaffolding" is put up, the termites appear to backfill this rough initial build, giving it structure (a process the entomologist is working to characterize).

But when and where do these bubbles start, and when do they stop? Paul's work actually may show that no chemical signal is needed—that the structure of the soil left by one termite is enough information for the next termite, to know what to do. Those cues, whatever they may be, are part of what the researchers hope to get at in these experiments.

"You ask how you come up with an algorithm? That's how you come up with an algorithm," Paul said. "You have to understand what the individuals are doing, how they interact with each other and how they interact with the environment, on that individual level—that's how you can build the rest of it. So everything you see above it comes from those initial steps; if you could predict what a termite would do in any given situation that has soil in its mouth or needs soil, you could actually create the whole structure."

Each of those simple rules in a termite's brain interacts with other rules to create the structures we see in the mound. And they often act competitively—and even at cross-purposes, the scientists are quick to point out. It is out of this system of competing agents that an order begins to arise.

Paul points to a perfect example: termite tunneling. And here, it's worth setting up a contrast with their archenemies, the ants. An ant that is looking for a termite mound will proceed in a "random walk," going back and forth and covering a lot of area. When it finds its target, it then heads straight home, to tell other ants where to go (and thus, save them the search). It's a very efficient way to forage. A termite, however, will have to forage very differently. After all, if they tunnel through soil in a "random walk" to forage, then they'll have to trace the entire meandering path back. That's not efficient. Their other option is to tunnel straight home—which is energetically very costly. But, if you balance the two needs at the same time—an efficient search pattern, and a speedy way home—the pattern that emerges is a branching network, just like the ones seen inside the Macrotermes mounds.

"It's actually optimized for search and transport, if you balance the two at the same time," Paul said.

Another example: Individual termites who are heading toward the tunnel are motivated to dig—to pick up a bit of soil and transport it elsewhere. Those who are coming back with a mouthful of dirt tend to deposit their load. Say that one termite digs up a mouthful of dirt while, right next to that hole, another termite traveling in the opposite direction deposits its load of soil. The termites that follow them will each dig deeper or build higher—and the end result of this asymmetry is that the tunnel grows into a diagonal branch. The angle that they form is around 55 to 65 degrees—which just happens to be the optimal angle in a bifurcated branching network. The competition between the two termites actually leads to an optimal solution.

The scientists believe that these animals can take in a number of different environmental factors—moisture in the ground, soil type, wind turbulence—and build accordingly. It is, in many ways, the opposite of architecture.

"An architect solves all that in his head and then he takes that and puts it into material. And if he's done a really good job then yay, he saved a lot of energy in not building things wrong," Paul said. "These guys do the reverse. They have no simulation but they're constantly in contact with the environment, so although there's a lot of trial and error that goes on when they're building, when they are done they are completely as optimized as they could be because they have constant input from the environment."

This is what you would call an agent-based system. The idea is that if each agent, or individual, in a system works its hardest to achieve its own selfish goal, when it comes into contact with a competing agent, the optimal solution will emerge out of the competition between the two. This system works on every scale in nature, from the cells in our bodies to the competition between species. That's what's happening when one termite is digging soil while the other is depositing; it's what happens when ants and termites attack each other. An optimum—or at least, a balance—emerges between the two.

"It's the same way our brains make decisions when faced with uncertainty," Scott tells me.

This is just by way of example, however. In nature, there are far more than two agents trying to reach their goals at the same time. So a successful system divides each function between the agents, who can build and integrate them all into a coherent solution. That's why systems like the termite mound are called "emergent systems," and it's a concept that Rupert Soar wants to bring to architecture and construction. Right now, an architect has to solve for all of these different goals inside his or her head—where the kitchen needs to be in relation to the dining room, how the airflow and plumbing need to be designed, how much sunlight should come in, how much material is needed, and so on—and then puts them to paper and forgets about them. But architects are human; they can only solve for so many variables. On top of that, they're creating this plan long before construction ever starts, without a full understanding of how to build for the future building's particular environment.

If you could design an algorithmic program that set up agents for all the different requirements for a particular house or office, you could design and construct buildings the way that nature does. If your computer program was embedded in a robot, or multiple robots, it could actually play that program out in real time and in situ as it actually constructs the building, taking the changing environmental factors into account. Ultimately, you could have buildings that, like termite mounds, can reprogram themselves, with the "agents" that drove the building's construction actually responding and adapting to changing environmental conditions.

Most people think about the future home looking clean, all-white, sanitized, as if dirt and foodstains no longer exist. But Soar envisions a future "living" building, one that's messy and almost chaotic, but also intelligent, resilient, information-rich, and multifunctional—just like a termite mound. For now though, there's a vast gap separating that vision and this reality, between the basic science and practical applications. If they're going to bridge that gap, the researchers know they're going to have to keep digging for more clues.

Turner doesn't even flinch as a big yellow backhoe lunges toward him, giant shovel outstretched, before plunging its claw into the ground a few feet away. It pulls out a clawful of dirt and deposits the soil in a growing pile nearby. The biologist is peering at the growling machine's target: A six-foot tower of earth rising out of the tall grass in the yard behind this crocodile farm in Otjiwarongo. To the farm's owners, this termite mound is little more than a pest and an eyesore, one that must be demolished. For Turner and his colleagues, this presents an opportunity: A chance to get some fresh fungus combs, hidden deep within the nest. Call this, then, an exercise in multifunctionality.

The backhoe's arm inches closer to the mound with every shovelful. Finally, it stretches out and the claw rips through the structure, digging down and scooping dirt out from the ground before retracting. Andre Pitout, the burly Afrikaaner operating the backhoe, expertly takes off layer after layer until Scott signals for him to stop. The mound—and the nest a few feet beneath it—is exposed, as if we're looking at a cross-section a third of the way through. Turner climbs into the newly created pit, searching for his target. Hundreds of termites, many of whose bodies are an ombré pale brown fading to white, scrabble over the holes, as if assessing the damage.

In the past, the team has taken a termite mound and sliced off thin sections from top to bottom, taking a photo after every slice. When digitally reassembled, it has allowed them to create a three-dimensional model of the mound's internal structure in unprecedented detail. But today, Turner has one main goal in mind—collecting the fungus combs hidden deep in the nest, which provide as the termites' food source. That's because, as it turns out, the termite mounds may indeed have their own air-conditioning system—but it involves the fungus combs, and it works in a totally different way from the Eastgate mall.

Termites are not simply mere wood eaters. Their bodies can't digest the tough cellulose fibers they bring home. Instead, they carry it back in their bellies, poop it out, and feed it to specific colonies of fungus that they bring to the nest and grow deep in their homes. The fungus secretes enzymes that actually digest the food, and the fungi (and termites) eat that digested stuff.

The termites have a complex relationship with fungi. Most fungal species are enemies to the termite nest, because they covet the chewed-up wood fibers that the termites bring back home. If those fungi were allowed to spread through the mound, they would quickly ravage the nest.

The termites can't control which fungi enter the nest. Every mouthful of soil they carry in is actually filled with all kinds of spores, ready to germinate, but the termites only want one particular type, *Termitomyces,* to grow. This fungal species in this genus are slow-growers; when they produce the enzymes and digest the stuff that the termites bring back, they take a while to absorb the resulting simpler hemicellulose sugars. This gives the termites a chance to grab a bite as well. That way, both the termites and the fungus get a meal (though perhaps not as much of a meal as the fungus would like).

How do the termites keep other, more malevolent fungi from growing and taking over? Are they like frantic gardeners running around and snipping off any weeds that rise out of the fertile soil? Not so much. The termites have a way of nipping this growth in the bud: by controlling the moisture in the nest. This slow-growing fungus can grow in slightly drier conditions than its fast-growing peers. So the termites keep the "humidistat" just high enough for their friendly fungus, but too low for the other fungi. They do this in part by bailing wet soil out of the nest—yet another reason the termites move water along with dirt.

Keep in mind, even this isn't a perfectly amicable partnership. This is another agent-vs.-agent relationship, where each party is looking out for their own interests. The termites are stealing the fungi's food, and they're also keeping the moisture level low enough that the fungus can't mature and spread through the mound. And the fungus would like nothing more than to just take over the nest and mound and start reproducing like crazy. Occasionally, you will see mushrooms fruiting on top of a mound—a sign that some of the fungus, in spite of the restrictions placed upon it, has managed to make a break for it.

The fungal combs play an active role in maintaining the right moisture gradient, too—after all, they don't want any other competitors for resources, either—and the researchers think it might do so with its microscopic surface texture. That's the theory, anyway. And so that's why Scott's here, digging through the nest. There are many combs inside a nest, and each one is housed in its own personal chamber, largely sealed off from the others (except for small holes between the rooms that the termites can use to send messages or slip through).

Scott gets down into the pit with a pair of shovels and starts digging. Rupert soon joins him. It seems like backbreaking work for the scientists, who aren't as young as they were when they started this project—but they don't seem to mind. The sun beats down, and Paul joins in the fray. He's got another goal in mind: to collect the queen, who will be hunkered deep down in the middle of the nest.

Sweating silently, Scott and Rupert finally hit pay dirt. Scott gently pulls out a brainlike, yellow object, cupping it gently in his hand. He reaches up to let others touch it before turning back to dig more for. The fungal comb is incredibly light, and it weighs and feels as if it were made out of papier-mâché—which, he points out, is what it essentially is. The comb's texture looks almost familiar, sort of like the regular bumpiness that you'll feel on those coffee-cup holders you get to hold multiple drinks—but instead of being rough to the touch, it's remarkably smooth.

Scott's excavation is drawing a crowd; three towheaded kids watch, fascinated, reaching out to touch the comb. Adults soon follow, asking Scott for samples to take home and show others. At first, because they don't know to handle the comb with delicacy, the samples break apart in their hands. Luckily, there's plenty more in the nest. One woman, a teacher, asks Turner to come speak to her class, and he readily agrees.

Andre, who has dismounted from the backhoe to see what the fuss is all about, has turned into Scott's most formidable student, asking question after question, surprising and impressing the biologist. "So we're standing on a lot of information right now," Andre concludes, while perched on the edge of the hole he'd just dug. He watches a termite approach his fingers.

"Those soldiers will bite," Scott warns him.

"Well, we work with crocodiles—they also bite!" the big

man bellows before laughing. He ponders the fungus comb in his hand.

"What does it taste like?" he wonders out loud. And then—to my secret horror—he sticks his tongue out and licks it. (The verdict: it tastes like moldy paper towel.)

As they'd dug, Rupert had noticed that there was actually another spire, not too far off. Scott nodded. "I reckon there're two queens," he says, to Rupert's surprise. As we dig away at the mound, he turns out to be right—Paul carefully pulls out two enormous insects with soft, obscenely pulsating bodies the size of small bananas, which he carefully puts in tubes to take home.

For me, that's not the biggest surprise of this mound. Paul points at a smaller-looking termite frantically running around the exposed dirt. It's a totally different species of termite, sharing space with the mound's main occupants in the heart of the colony.

"You'll find ants in there, you'll find insects," Rupert says. "You'll find all sorts of organisms coexisting inside a structure like this—just like a city."

Staring at this multifunctional mound, I'm struck by Scott Turner's multifunctional work. Today's excavation is a demolition, even as it's an academic examination and also a public education. In that way, the scientist is a little like a termite himself, operating on multiple levels, perhaps without always thinking about it.

Back at the ranch, I find Rupert at the desk by his cot, hunched over a microscope, the lens trained on a small fragment of fungus comb. He takes a thin insulin needle and drops a tiny droplet of water onto the comb. He steps aside and I peek through the scope, squinting as I turn the knobs to adjust. And there it is, the water droplet still sitting perfectly on top of the surface. Eventually, this droplet will sink into the surface—but it can take more than an hour, Scott says.

"If you zoom right in and use the focus, you'll the see the hyphae, the superfine hairs that the fungus coats the entire structure with," Rupert says. "And those hairs are repelling that water. Can you see that water droplet that's floating on top like it's on wax? That's how repellent they are. But they can change from being super soaking—hydrophilic—to super hydrophobic."

This is how the fungus keeps its competitors out, Scott says. If the humidity goes above 80 percent, all those fast-growing, wood-rotting species can start to grow and will quickly take over the mound. So right around the 80 percent mark, the hyphae—this network of fibers—actually become hydrophilic, sucking in as much water vapor as possible to pull it out of the air. If the humidity is below 80 percent, the hyphae stay hydrophobic, blocking the water from being absorbed by the comb and thus keeping the humidity just high enough so that it can survive.

Scott tells Rupert to put a droplet right on the dark edge of the fungal comb, whose surface the hyphae hasn't yet fully covered. Zip—without the hyphae as a barrier, the droplet sinks in immediately, as if sucked up by a paper towel, which is essentially what the comb is made of.

Understanding this microscopic surface, and its ability to change behavior according to the environment, could be key to creating buildings free of air-conditioning units. Air conditioners work in large part by pulling moisture out of the air, reducing the humidity. If you were to create a passive structure that could do this—perhaps by coating the walls with a material that had these hyphae-like properties—you would accomplish much of what AC does without having to constantly recirculate the air, waste precious energy, and pollute the environment.

For now, though, how the fungal comb manages this trick is largely unknown. The comb seems to have a number of properties that defy intuition—for example, it actually seems to be more brittle when wet than when dry. A lot more basic research will need to be done on the fungal combs before such passively air-conditioning walls are possible, and commercially viable.

Hunter has come into the room and taken a turn at the microscope. He leans over the eyepiece, feeding droplets onto the surface with the insulin needle.

"Is it possible to take this material that the hyphae is made out of and then homogenize it and make a flat surface out of it?" he asks. Different materials can repel water for different reasons. Some have a particular physical structure that allows them to repel droplets; for others, it's the chemicals they're made of. If you could make a flat surface out of this stuff and still have the same effect, then it would mean that the hyphae's powers lie largely in its composition.

"Don't know," Scott replies. "Maybe. We don't actually know why the fungal hyphae are hydrophobic. We don't know if there's surface sculpting, we don't know if there's some kind of protein in there—it could be both possibilities."

A couple days later, after hiking up the technicolor lichen-painted cliffs of the Waterberg, Rupert takes Hunter, Sam, Max, and me to a ranch in Omatjene, which lies about an hour west of Otjiwarongo. This is where Scott and Rupert performed many of their early experiments, several years back. Officially, we're here to check on some long-stored supplies, but it's more or less an opportunity to see the plaster cast of a Macrotermes mound.

It's a sunny afternoon; our car slowly crawls past herds of goats through calf-high grass, until we stop and get out near a large caged structure. A few steps through the dry brush and we can see it: An enormous white, rounded cone, higher than our heads, filling the cage. The central tunnel and its main arteries, thick and strong, branch out toward the mound's walls in ever-thinner passageways, until they reach the egress complex, the dense tangled mess of holes that give the mound's surface its porosity. It's vaguely reminiscent of a map of interconnected freeways, with the egress complex being the exits onto interwoven city streets.

Rupert and Scott had managed to pour gypsum into the top of the termite mound, allowing it to fill every nook and cranny. Once it dried, they carefully washed the soil away over the course of several weeks. This is what remained: an incredibly intricate network of passageways that reminded me of the blood vessels in our own bodies, or the nodal structure of the cosmos.

One side of the mound seems to be missing; apparently someone on the grounds must have left a gate on the cage open, and the cows had come and nibbled at the plaster for the salt it contained.

Rupert points to the network of surface conduits and the smaller passageways that branch off into the egress complex. That network of small and large tunnels is key to the mound's ability to take in fresh air, said the engineer, who has taken earphones and tapped on the tunnels, listening to the different frequencies.

"I can play these tubes like a glockenspiel," he says of the hollow tunnels.

"Air is always acting transiently on this," he adds. "This is a series of organ pipes and all with different resonant points, lengths." As the air sloshes back and forth at different speeds, it can hit these tunnels at the right frequencies for them to resonate, forcing the stale air inside to mix with any fresh stuff that has filtered in through the semi-porous surface. "That creates this gradient by which oxygen and carbon dioxide are exchanged, or gradually moved from underground, from the nest, right up to the surface skin and then away."

There's something strangely lopsided about the mound—the structure of the egress complex seems to be more delicate, and better kept, on the north side than it does on the south. I point it out to Rupert. He agrees. The termites tend to build toward the sun, which is why the egress complex and the finer-detailed tubes appear on the north side, where they're likely to get all-day sunshine. The complex on the south side is often neglected, sometimes washed away and never fully repaired.

Farmers and other locals would come to see the plaster mound—while termites are a common sight in Namibia, it's not often you get to see the internal structure of one. When they saw it, Rupert said, the response was very often religious; viewers would drop to their knees, reciting a verse from the Bible: "Go to the ant, thou sluggard; consider her ways and be wise."

Seeing the mound, made of so many interconnections that it almost looks like bony gossamer, I can understand why. The brain sees in it a beauty—but one whose pattern is indecipherable, just beyond reach, so the design wobbles in the mind's eye between order and chaos. It's hard to look away.

While the physicists or engineers in the group don't seem quite ready to kneel, they're clearly fascinated. They gaze up and up along the interlaced tunnels toward the very top, lips parted but silent.

Rupert muses on another theory of his as he and Max look at the mound. The structure is shot through with tunnels at different scales; there's almost as much air as there is dirt. How do the termites know how to dig their holes without causing the whole thing to collapse? When do they start, and why do they stop? Rupert thinks the termites might not just be responding to moisture, or to wind, but also to sound. If you were to rhythmically flick a material, you might notice that the frequency of the sound rises as you put pressure on it. It's possible that termites can pick up on these vibrations as they trundle through the tunnels, giving them a sense of whether it's best to excavate or reinforce a given structure.

"If you look at the work Paul's done with mining termites in wood," he says, "termites can track each other and can make the tunnels join precisely there. So are these termites doing something similar?"

If they are, then perhaps there's some kind of resonant template that lets termites get feedback from the vibration of the material around them, allowing them to build up when necessary, and dig away at parts of the structure that are no longer needed, as if they were removing the mound's built-in scaffolding. They've certainly seen termites tapping at the mound—but why they're doing so remains a mystery.

"It's confusing data and I'm not seeing anything obvious. I want to take those recordings I've got now and play them back to termites in petri dishes." That's what he had been hoping to do at some point during this trip, he says. "There's just no time. Another project, another time."

Rupert imagines building a structure like the Eiffel Tower like this, using sensors and a microphone, to build the structure adaptively, by tracking how much load each beam is experiencing through the change in their vibrating frequencies. Using this feedback, you could actually remove beams during the building process as they no longer remain useful, thus creating a sturdy structure with the least amount of material necessary. Rather than trying to plan out the most efficient use of supplies beforehand—the way an architect would—you find the solution as you go.

"You're building that process down tighter and tighter, almost to the point where you're building the fabricator building the car," he says to Max. "You're squeezing that timeline, until it's almost immediate. To now!"

It's an idea that echoes a critique that Rupert has long held about the current building industry. Architecture is often treated as a completely separate process from engineering and construction. The architect responds to the client's desires, creates an inspiring (and perhaps impractical) model based more on form rather than function, and for the most part avoids interacting with engineering and construction. He doesn't know the cost or how it will be built, and he doesn't work with the builders to actually modify his design based on their feedback. Other industries—automotive, aerospace, medical—have integrated these steps, and have done so digitally, thanks to a discipline known as systems engineering. The building industry has not quite gotten there yet—and still it's resisting, Rupert adds.

That approach is in complete opposition to the way termites build, and where Rupert and Scott see the future of architecture. To the termites—and thus, to the researchers—a building isn't a fixed object; it's a process. Rupert sees a far greater potential for future homes, if they were to harness the algorithms that drive the termites' adaptive, multifunctional building processes. Modified for human homes and needs, you could take the same construction algorithm and apply it in three different environments—in a warm lowland, a frigid hillside, or by the ocean—and get three totally different buildings.

But that wouldn't be the end of the construction process. That could be ongoing, for as long as the building stands, as the climate and the inhabitants' needs change. After all, termite mounds are constantly changing, built up and torn down, whether by rain, wind, or the termites themselves.

"It's hard to nail down what's cause and what's effect in these systems," Scott had told me earlier. "And the key to me as a biologist is that this is very dynamic process. There's no such thing as an object in biology. These are all transient types of phenomena.

Matter is flowing in, matter is flowing out, the actual composition varies with time, even though the form might be the same, and that's where architecture hasn't quite caught up yet because they don't know how to build a dynamic building in the way that a living system is dynamic."

Such industrial changes are probably a long way off, though. There's still so much basic research left to do. Even when it comes to airflow in the mound—the aspect of termite construction that Scott and Rupert have most focused on—they're still not sure how it works.

"We still are trying to pin down the absolute mechanism here," Rupert says. "We understand the egress complex and how gas exchange occurs through it, but we don't understand this kind of macrostructure. And that's what Sam and Hunter are doing. To bring the most sensitive devices in the world. Because we've tried everything in the past and we just can't measure this transient system."

Sam and Hunter have been examining the connectivity, trying to understand the ways that air could move through the mound. As we make to leave, Sam calls out, as always a little louder than necessary: "Nonscientific question. Would Omatjene begrudge me if I peed in their bushes?"

"Fancy having a Harvard graduate peeing on your farm! They'd be delighted," Rupert says. "To future generations: the famous Nobel laureate Sam Ocko, he peed right there."

On one of the last nights at the ranch, I follow Hunter and Sam through the dark as they go to check on the mounds at night. I look up to see if I can spot the Milky Way, but it's no use—the full moon is so bright that it easily illuminates the path in front of us. Hunter asks Sam to turn the hand lantern off; he likes walking in the moonlight. Nights on the ranch are not quiet. Unseen cattle bray, sounding much closer (and more ominous) in the quasi-dark than they do in daytime, and I can't help but start to think of all the threats that could sneak up unannounced: a territorial bull, a troop of baboons, a cranky cheetah. I shiver and hurry to catch up.

After what feels like a very long walk, and after I've lost all sense of direction, we finally arrive at the mound they're looking to study. The physicists want to get readings on the mounds at all hours of the day (and night) so that they can see if there are any differences in the air movements in the mound. Their sensor has two modes: One to detect steady flows of air through the passageways, and one to detect transient flows—the kinds that should be triggered when transient winds hit the outside of the mound. Sam pulls out his stool and laptop, Hunter readies his laptop, and they get to work.

The two physicists have easily become my favorite scientific odd-couple on the ranch. I could not imagine two more different people in personality and bearing: Hunter is slight and soft-spoken, careful to think before he speaks, and keenly aware of noises and other stimuli around him; Sam is taller and louder (though he doesn't seem to realize it), is easily distracted by data, and has an endearing habit of knocking into things. Every interaction between them is hilarious, in that muted, sweetly quirky tradition of TV comedies like *Parks and Recreation*. They're also the newest additions to the group; their first trip with Scott and Rupert was just a couple months before this, to study the mounds in India.

Compared to the ones in India, the mounds outside the ranch are difficult to work with—crumbly and often dead, probably in part because of the extended dry season Namibia had been experiencing. It has made gathering good data difficult for the physicists. But Hunter has promised to show me some of the data they gathered from India, to give me a sense of how the information they assemble gives them a picture of the mound's dynamics.

After testing a few more mounds, we head back to our sleeping quarters in a side building. Hunter puts the equipment away while Sam pulls out the computer. First, they show me a termite mound from the Indian species, *Odontotermes obesus*.

"Whoa—they look like castles!" I say. The mounds are pointy and seem to have fluting wings coming off of their craggy surfaces that in some places look like leaning buttresses, and in other places look like separate, smaller towers, giving it a fairy-castle-like effect.

The physicists measured air currents inside the mound, and they quickly realized something that did not fit with Scott Turner's theory: They could not detect any transients. The surface walls were impermeable—there was no way for transient winds to penetrate. On top of that, the pair do find that air seems to be circulating in the mound—which also appears counter to what Turner thinks is happening in the Namibian termite mounds.

"In this experiment there were no transients that we measured; it was just a steady flow, turning in one direction, turning in the other direction," Hunter says.

So do Turner's theories not apply to the Indian species? Hunter avoids getting involved in the biological debate and just sticks with the numbers.

"We haven't addressed at all any biological thing of what it's trying to do, this is just what the flow does and what drives it," he says.

The scientists measured the temperature of the air in the flute-like appendages, as well as in the mound's central cavity. They found that as the day warmed, the smaller volume of air mass in the fluting areas would warm up faster than the larger air mass in the center; this temperature difference would power the circulation going in one direction. At night, the flutes cooled down faster, thus forcing the circulatory system to switch direction. As the circulating air neared the mound's surface, it would be able to exchange gases with the outside air through the walls. So while the system circulates air through heating, it has nothing to do with the termites' collective metabolism, and everything to do with flushing carbon dioxide out of the mound. So, even with this very different style of mound, Scott's conclusions at least partly apply.

"This is what Scott's been trying to say for a long time, too," Hunter says.

I feel some hesitancy coming from Hunter when he describes the results; perhaps no one likes to be the bearer of less-than-ideal news, no matter how impartial the data or neutral the messenger. Still, neither Turner nor Soar seem anything other than interested in the results; perhaps it simply offers up for study a new mechanism with unexpected practical applications. And for the moment, not even Hunter and Sam seem to know exactly what's happening in the Macrotermes mounds in Africa.

It had seemed a shame to leave southern Africa without seeing the Eastgate Center, which lies in the heart of Zimbabwe, just two countries over from Namibia. Debunked or not, Eastgate is possibly the most well-known "bioinspired" modern building, and something of a poster child for biomimicry in architecture. Mick Pearce, the architect, said he'd gladly show me around his famous work, and so I boarded a plane to Harare.

The Eastgate Center works by taking advantage of the day-night cycle. During the day, giant fans placed at the bottom of the structure circulate air through the floors, using cool concrete teeth to sap the heat out as it flows past. The air rises through the building and out through chimneys, thanks to the stack effect. Over the day, that cold concrete slowly warms up; and so at night, the fans blow cold air past the blocks at a high rate to cool the concrete down for the next day.

The main critique Scott and Rupert have of the Eastgate Center is this: While it seems to work just fine, it was based in part on some fundamentally misguided scientific theories. After all, if the idea were truly based off of what termites do, then the building shouldn't require giant fans to constantly be pushing the air upward; the system should be able to refresh the air passively, as a termite mound does. If you have to make significant changes to a supposedly bioinspired design, then perhaps the design wasn't accurate in the first place. And of course, as Scott Turner has shown, there is no apparent stack effect, and the purpose of the termite mound's design is for respiration, not temperature control.

Scott and Rupert, in spite of their critique of the Eastgate Center, seem to have a healthy liking for Pearce. Perhaps it's because they've found in Pearce something of a like mind. The architect is fascinated by nature; his shelves are filled with a number of biology texts, including some written by Scott.

When I arrive, in fact, the first place he takes me is not to the shopping center, but to the termite mounds dotting a golf course about ten to fifteen minutes outside of his home in Harare. It's quite early, around 6:30 a.m., and a mist still hangs over the rolling field. Mick unleashes his two young but large black dogs—Jack and Banjo—and they go bounding into the fog, leaving us to tramp through the damp, knee-high grasses alone.

Walking stick in hand, Pearce, who is in his seventies, heads toward an overgrown spot, a sort of three-meter-wide circle where the grass towers several feet overhead. He dives in, hands parting the tall thick reeds, until I almost lose sight of his blue jacket. I grit my teeth and follow him, plunging toward the center.

He stops in the middle, and I run into him. I look around—is there supposed to be a termite mound here?

Pearce points to the ground. There, in the center, is the tiniest termite mound I have yet to see on my trip: A structure less than eighteen inches tall that isn't a mound, but kind of a fat column—as if you took a rainboot and buried the foot in the dirt, leaving only the shaft above ground. I stick my hand wrist-deep into the hole, which seems to reach almost all the way to ground level. The interior is several degrees warmer than the surrounding air, and damp like dog's breath—I can feel the moisture sticking to my skin.

"Oh wow. It is warm! It's like a sauna in there," I say, totally surprised. This, honestly, is not what I had expected to see: a completely different-looking termite mound, squat towers almost as alien to the Namibian mounds as the Indian fairy castles were.

"This argument I had a little bit with Scott," Pearce says. "When the grass is this tall, there can be no wind action at all on this tower."

I see his point. It was hard enough for us to push though those grasses; I'm sure it would be an incredibly effective wind dampener. If these were the mounds Pearce saw when he was pondering the stack effect as a "bioinspired" process, he could be forgiven for thinking so.

These oases of head-high grass punctuate the yellow-gray field; we hop over to another, and then the next, finding a mound in the middle each time. One of them even looks like it's steaming, and I can't help but think of my own breath fogging up on chillier California winter days. We run into one of Pearce's neighbors and his wife, out for their morning walk—a former potato farmer, "chucked off his land" by the government of Robert Mugabe, the architect tells me as an aside.

It's May in Zimbabwe, and approaching what passes for the

dead of winter, and so there's little activity visible on the mound's surface. If this were summer before the rains hit, he says, the mound would be abuzz in activity. Pearce actually pulls a photo up on his phone, and he's right; the interior of the tower is absolutely covered in little off-white bodies. I'd have been far less willing to stick my hand into that.

"In the rains, just before the rains, that's when it all happens—when this place is alive with termites," he says. "Their whole cycle is linked to the seasonal cycle. If you see them every day, to see what's going on, you get a picture, which may be different from what Scott and Rupert see, because they can't be there all the time."

But perhaps these mounds are also able to take advantage of wind energy, when the situation calls for it. Often, golfers (with presumably bad aim) will burn the fields to get rid of the grass so they can find their lost balls more easily. Stripped bare of their grassy enclosures, the mounds' shapes seem to change, Pearce says; the lip of the hollow tower actually seems to grow a tip, extending toward the wind.

The architect hates it when they burn the grass, though.

"They're denuding the ground," he says disapprovingly, as we walk along. "They don't understand that plants actually die and the residue goes back into the soil. If you drag it off, it ends up somewhere else. Usually in the air. There's far too much carbon in the air and not in the ground."

Pearce thinks a lot about how man-made structures interface with nature. Currently, he's thinking a lot about how we interface with the water systems in our environment—usually blocking or diverting them, rather than facilitating them.

"The problem with cities is you create huge areas of sealed surface and tarmac roads so when it rains, the water runs off into the rivers and down to the sea," he points out. "So then to make the city work, you build dams, above the city. It's really a bit crazy. What I'm trying to do is to get city planners to get the water back into the ground."

He lifts his stick and points to a large nearby tree.

"Most architects seem to be wildly enthusiastic about form, pure form, and that's what drives their design; I don't like to see that," he says. "I'm interested in process. That inspires my work. So when I look at a tree, there's a very beautiful tree. And I love its form and shape. But in fact, another way of looking at it is as a bridge between groundwater and the clouds. So it's a process. And it's changing all the time. It's an adaptive structure."

Now, he says, he's designing buildings that are inspired by the "process" of trees—buildings that capture water instead of shedding it.

Processes, adaptive structures—the researchers in Namibia may have some differences in thinking with Pearce, but it's easy to see why they've thought of working together. The more Pearce talks about his work, the more he sounds like Scott.

We head to Eastgate later that day. For this building, Pearce did not limit his inspiration to termite mounds. Along with the trellislike geometry, bits of the building jut out in a three-dimensional zigzag pattern, giving the surface a sort of accordionlike feel. Pearce calls it "prickly," because it's a design he says was actually inspired by the barrel cactus. The cactus uses its ridged, prickly surface to minimize the amount of sunlight that hits the surface, and thus, the amount of heat absorbed.

"If you put a prickly thing in the sunlight you'll see that it has a lot more shadows than the smooth thing," he explains. That means that while the tips get hot, the rest of the building doesn't. And at night, the zigzagging face actually provides more surface area than a smooth face would do to radiate heat, which allows them to cool the building faster. It's a win-win solution, both for the cactus and the building.

Mick Pearce did not set out to build a climate-controlled, AC-free building just for kicks. The client that owned the property, life insurance firm Old Mutual, wanted something without air-conditioning in order to keep the costs of constructing and then maintaining the building as low as possible. They also wanted to use local resources, because imports were extremely expensive. That's when he cast about, looking for ideas on how to control temperature.

"I was afraid to mention termites to my clients, afraid they'd chuck me out of the room," he says.

The inside of the center feels surprisingly light and airy; the trellislike concrete designs continue inside, and a public street actually passes right through the middle of the complex, with just a glass atrium above to protect against the elements. The whole building has a blue-green feel from the metal elements like the elevators, which are hung above the public street, so that they're ventilated by the outside air and not the building's internal system.

Pearce takes us up to the eighth floor, where his architecture firm's offices are, to show me what the individual suites look like, and how they connect to the larger system. Down at the very bottom of the building, giant fans pull cold air in at a really fast rate at night—turning all the air in the building over about ten times each hour. All that cold air runs through ducts beneath each floor containing concrete teeth, pulling all of the day's heat out of the structure and carrying it away. Then, during the day, the fans blow air through the ventilation system at a much slower rate, about twice an hour, so that the cold concrete teeth can now cool down the passing air before it enters each room. Pearce points to the ceiling, where there are a series of basketball-sized holes; once the cool air comes into the room from the floor, it warms, rises, and then escapes through those holes, into the ventilation shaft, where it floats up and out through the chimneys. Popular estimates put the building's energy savings at 90 percent, although it should be said that Harare is blessed with a relatively moderate, manageable climate.

Termite ventilation may not work in the same way, or even for the same reason, that the Macrotermes mounds in Namibia do. But when Scott looked at the Eastgate design, he noticed something else: Pearce, in order to make his design work, had had to put large concrete blocks into the foundation as a heat sink. This wasn't inspired by what he saw in termites; this was an architect's solution to a design problem. But as it turns out, this actually is how Macrotermes mounds shed heat, Scott says—by using the soil as an enormous heat sink.

By dint of being a good architect, Pearce and his colleagues "converged on the same solution without really realizing that they were doing something that termites actually do, but no one actually knew about yet," Turner said.

Pearce points me to the window of his office, where I see a number of low-lying connected buildings across the street. That, he says, is his next project.

The Eastgate Center is facing an economic problem: it has roughly a 50 percent occupancy rate, Pearce tells me. Meanwhile, the informal markets across the street have begun to thrive, as squatters move into the building and set up shop to sell their wares illegally, without having to pay rent. It's an issue reflected in the larger Zimbabwean economy, thrust into particular turmoil around the year 2000 when Robert Mugabe began seizing, sometimes violently, the lands of white farmers and redistributing them to black residents, without taking into account the loss of those white farmers' institutional memory and experience. Hyperinflation became so extreme that the country abandoned its own currency and switched to the U.S. dollar in 2009. As instability and crisis have grown, people have pulled their money out of the formal sector and been increasingly doing business in the informal sector. Zimbabwe's informal sector is valued at about $7.5 billion—mind-boggling for a southern African nation whose 2013 gross domestic product was just under $13.5 billion.

"After independence here we built a lot of shopping centers modeled on California, pretty much," Pearce says. "Well, they're now empty. There's nothing going on in them. Or if they are, they're losing money. And outside the shops, on the pavements, people are selling and trading. They find it a lot cheaper because they don't have to pay any rent or any tax."

Pearce has some ideas about adapting Eastgate to the changing times; he wants to split the offices, make them into smaller stalls, to reduce rents and accommodate the style of sellers in the bazaarlike markets. He also wants the owners to consider turning some of the suites into residential apartments, so that the building can be used day and night, and which may help revitalize the area, reducing violence and bringing in business.

If you have people living in the building, "then you get street cafés, nightlife—not only that, but you can pay for your infrastructure because it's so expensive," he says.

But he saw a larger opportunity when he heard that the owner of the property across the way—which also owns the Eastgate Center—was looking to take back the property from the current, unlawful occupants. The group works by charging sellers for a position—usually a table, about one meter by two meters, a dollar per square meter—where they can set up shop. It suits the sellers well enough; it's cheaper than paying rent in a center like Eastgate.

"No one's allowed on that site at the moment; it's run by a bunch of crazy crooks, actually," Pearce comments. "They call them 'war vets.'"

Outside, people also hawk wares on the street: a woman selling vegetables who carefully balances two baskets of tomatoes on top of each other; several ramshackle storefronts for Eco-Cash, a cell phone–based money transfer service; and a parked compact car with the passenger door wide open and a red cardboard sign that says SIM REGISTRATION DONE HERE taped to the windshield.

The property owner wants to turn all of this into a parking lot. But Pearce doesn't think anyone wants or needs another parking lot in the area; after all, many of downtown's daytime inhabitants walk or bus into town. Instead, he persuaded the owners to let him try a different kind of termite-inspired project: one that worked with the sellers in the informal market, rather than against them.

Pearce's plan is this: to turn the buildings into a huge open structure that allows them to set up stalls that they'd rent out to the sellers. His design would offer them much better conditions, including places to store their possessions at night. It would not be completely free-form—he has plans for a food court on the building's second story—but in large part, the sellers would be able to carve their own niches out of the space, and their collective activity would determine the overall layout of the market. The sellers, then, would be acting rather like termites, unwittingly creating an emergent market system that is flexible, able to adapt according to the changing needs of buyers and sellers.

"I'm just providing a framework," he says. "So it's sort of an intermediate effect, where the occupant is the agent of change in the building."

With precious few resources—some dirt, and its own sticky spit—the termite manages to build an entire towering city, and adjust it according to the environment and to its own needs. And perhaps humans are not so different: ever resourceful, unwittingly part of a system greater than themselves. The more Pearce talks about the informal economy, the more I begin to see the similarities.

"I'm fascinated by the way people build their own cities here," Pearce told me. "In a huge, massive way, out of nothing. Out of bits of tin and recycled plastic. And they create their own economy."

It reminds me of something else Scott said, back in Otjiwarongo, about the future of the building industry.

"That is the interesting dilemma it seems to me that architects are grappling with. Is the architect a specifier of living places . . . or are these people facilitators of people deciding what they want?" he asked.

Pearce, for his part, seems more than ready to choose the latter, or at least forge a solution at the midpoint between those two options. It strikes me as odd, simply because I would assume that an architect would like to preserve his or her elevated and insulated perch in the industry.

But it seems Pearce has never quite been like that. Rupert Soar, for all his critiques of Eastgate's design from the standpoint of biomimicry, actually may work with Pearce in the future—perhaps because their visions of what the future of architecture and construction should look like are not all that different.

Soar believes that, rather than operating separately, future architects, engineers, and builders must work together during the construction process. Digital construction will help break down barriers between those disciplines and allow them to act as an integrated whole, as nature does. It's the principle behind much of Paul Bardunias's work, why the tunnels branch at the angle that they do, a compromise between the termites who dig and the termites who build, a compromise between searching for food and heading quickly back to the nest. If architects are one agent, with certain design goals in mind, and the builders are another agent, with materials efficiency in mind, then the two need to work at the same time, each pushing for their own objectives, to build the best building that they can.

While building the Eastgate Center, Pearce in many ways did just that—he had daily meetings with the construction company, and worked with them when they ran into problems to come up with a solution. So while the building itself may or may not be all that termite-like, perhaps the process was more like the insects' building process than the architect himself realized.

Pearce, Soar says, was thinking like a termite.

6

HIVE MIND

How Ants' Collective Intelligence Might Change the Networks We Build

Few creatures you might find on land look as humble as the slime mold *Physarum polycephalum*. The bright yellow mass lives in moist damp areas, such as the soil in your well-fertilized garden or under a rotting log in the woods. The slime mold is not technically a mold—it's neither plant, animal, nor fungi—and so is often lumped in with the protists, an unofficial and unrelated grab bag of critters that are single-celled, either on their own or in colonies that blur the distinction between what makes a cell an organism and what makes it a mere unit in a larger living body.

While the slime mold does not live up to the second half of its name, it certainly fits the first. With a sickly wet shine, it looks like radioactive snot shot from the nose of a giant. It can take many forms: translucent and membranous, or balled up like a gelatinous missile. It has no arms, no legs, no nose, no eyes, no brain—no apparent appendages and abilities of any sort. And yet, with enough food to fuel its growth, it can double its surface area in a day. If it can't seem to find anything to nosh on, the amoeba can extend footlike tendrils of its body, called pseudopods, and start crawling to a more suitable location.

Its strange talents go well beyond these physical abilities. In 2010, Japanese scientists placed some slime mold in the center of a surface laid out like a map of Tokyo. At each of the surrounding cities, they placed a meal of oats, a favorite for the slime mold, which in the wild dines on bacteria and fungi found in forest debris. Slowly, inexorably, the slime mold expanded its reach, growing until it made contact with the oat oases and then self-pruning until only the connections to each food source remained. The result, in the end, looked very much like a map of the Tokyo rail system.

Let me put into perspective how incredible the slime mold's achievement really is. For one thing, a transportation network has to juggle a few different goals: It has to allow people to reach their destination as quickly as possible, which means not having to take five stops just to get to a city that's next door. But it also has to do so without building too much infrastructure, because government funding isn't unlimited. Finally, it has to be fault-tolerant—with enough built-in redundancy to withstand breakdowns in the network and reroute traffic around blockages.

It takes whole teams of experts with advanced degrees to build a system like this. It takes years of careful planning and oversight. And yet the slime mold manages it without a single neuron in its amorphous body, perfectly optimizing the network it makes between oat-filled hubs to minimize the resources used, maximize the oats it reaches, and to be resilient enough to withstand the occasional breakage in the network.

The slime mold's abilities don't stop there. The amoeboid is also very good at solving mazes, filling every nook and cranny with its body until it finds the oats at the other end and then shrinking back only to the very shortest path. Researchers have even found that it seems to be able to anticipate environmental changes, essentially flinching when a light flicks on at regular intervals (the slime molds prefer the dark). And in a recent study I wrote about for the *Los Angeles Times*, they even seemed to exhibit a primitive form of intelligence, able to "learn" that a bridge laced with bitter-tasting quinine was actually safe to cross.

Don't worry, this chapter will not be all about slime molds, even though they really are endlessly fascinating creatures that could overturn a lot of our preconceptions of intelligence, adaptation, and resilience. But because this creature it is so primitive, it serves as a perfect example of the idea of collective intelligence. Slime molds basically spend their lives as one big cell—but inside, they contain a multitude of nuclei. It's thought that they're formed when individual cells with flagella for tails swarm together and formed the giant cell we see, losing their individual identities in the slime for the sake of overall survival. It's not a dissimilar philosophy from that of other swarming organisms, including the worker bee, which must essentially disembowel itself to sting an intruder, or a worker ant that takes care of the queen's brood, never being able to lay eggs of her own.

All of these creatures display what's known as collective intelligence. It's an idea that holds endless fascination for humans, given that we see ourselves as such an individualistic species. We're used to the idea of building structures and designing networks with some kind of a central architect—just think of the 2003 movie the *Matrix Reloaded*, the second film in the Matrix trilogy where Neo meets the Architect, who apparently designed the eponymous system. But slime molds don't need brains; ant queens aren't urban planners. Nobody is at the helm in natural systems, and yet they manage just fine, having crawled across the face of the earth many millions of years before the first human stood on two feet.

Instead, these incredible, adaptable, potentially intelligent, and extremely resilient systems are the result of simple interactions between one part of the cell (or one animal in a colony) and the next. And the slightest of tweaks in those very simple interactions is what allows the complex behaviors of these systems to emerge. It's a way of designing networks that researchers are really just beginning to explore.

Deborah Gordon's lab is crawling with ants. They live in a large box that's the size of an overstuffed suitcase with a red-tinted, transparent lid. This isn't mood lighting—some species of ants, like the *Pogonomyrmex barbatus* underneath the glass, can't see red light. This allows the Stanford University biologist and her students to check on the insects, more commonly known as harvester ants, while allowing them to think they're safely indoors, in the dark.

Inside the large box lies a constellation of clear small boxes, roughly the size of a smartphone, each connected to the others via clear plastic tubes. This is the nest where the ants spend their days, bringing in food and tending to the young. Some of them are damp, painted with wet plaster so that water forms in droplets on the ceilings that the ants then collect and drink. One tube opens not into another little box, but onto the outside, like a sewage pipe in a cliffside hanging over the ocean. It's outside of one of these that I can see a collection of dark, crumpled forms—likely a combination of the ants' midden, where they dump their trash, and their cemetery, where they leave their dead. Gordon makes sure the ants are well fed with seeds, but just in case any ambitious workers try to climb their way out of the premises, she's painted the walls with a nonstick coating akin to Teflon.

It turns out that these small boxes are actually the lab's first-class suites—roomy, featuring plenty of amenities. Farther into the room are the economy lodgings—a row of smaller containers whose sides are transparent instead of red and hardly larger than a shoebox.

"We're basically just [keeping] them to get their DNA, so they won't get such a cushy life," Gordon explains.

But modest as they are, these colonies seem to have something the larger one doesn't: names. On one box I see "Bertha" written in all caps; on another, "Artemis." Whichever student finds the queen when they dig up the colony, Gordon tells me, gets to name it. Given that essentially all the ants in the colony are female, the women's names seem appropriate.

Gordon opens the door back onto the hallway, bright from the afternoon sunlight streaming in from the windows at one end. As in most campus buildings, the doors are flanked by cork bulletin boards and taped-up presentation posters. On one side of her office door, which sits almost directly across from the lab, hangs an enormous, chaotic black-and-white poster of a scene teeming with slightly anthropomorphic ants, rats, octopuses, a leopard, and even a snail with a clipboard.

The inside of her office is chock-full of ant art: a giant ant the size of a Chihuahua, made of three large rocks for the body and metal rods for the legs and antennae; sketches of the insects; a photograph of a dog pondering what to do with an insect scurrying across the floor beneath its nose. The table is covered with a dozen toy ants of various sizes and colors, from the size of a golf ball to a soccer ball.

Deborah Gordon did not set out to study ants. After getting her bachelor's from Oberlin College in French (with high honors), she got her master's in biology at Stanford University, where she wrote a thesis examining whether catfish really did, as reported, jump out of the water right before earthquakes struck. The hope had been to try and figure out whether this phenomenon really existed, and what the mechanism was. Gordon wanted to see if catfish responded when seismic activity changed the electric field of the water, and they did. There was just one problem.

"Every time there was an earthquake, the catfish all died," Gordon said, before adding: "It was very statistically significant, but also inexplicable."

A professor at Duke University piqued Gordon's interest in developmental biology, and in studying the development of embryos. But this was the late 1970s, and many of the sophisticated imaging techniques that scientists use today to watch living cells moving and interacting did not yet exist. So instead, the biologist turned to a system whose components she could see—ant colonies.

"I was interested in systems without central control. So an embryo is analogous to an ant colony in that nothing directs the development of different cells into different kinds of tissues, it happens through interactions among cells," she explained. "And I was wanting to look at a process like that, that I could actually see because I learn best by seeing. And it's difficult to watch an embryo as it develops but it's possible to watch ants interacting with each other and to see how their interactions lead to the behavior of the colony."

The idea of the colony as a sort of superorganism had been taking hold in scientific circles, and researchers were starting to draw parallels between the behavior of swarming organisms and the behavior of cells in an organ. Four decades later, the many intricacies of ant society continue to hold her in their thrall. I ask her if she had ever considered switching fields again, turning to study something else.

"I have thought about it a few times along the way," she said. "But there's always been something I wanted to know about ants that took priority."

The scientist hasn't suffered from a lack of questions. Gordon has traveled from the jungle to the desert to study an array of different ant species, but it's the harvester ants in the southwestern United States she knows best. She heads down to Arizona each summer to check on the same spot she's been studying for about three decades, filled with hundreds of colonies. Worker ants generally only live a year or so, but their queens have been known to live twenty-five years in the wild, perhaps more—and colonies live as long as their queen does. This means that Gordon has seen some of these colonies from their scrappy youth to their illustrious golden years.

This particular species of harvester ant, known commonly as the red harvester ant, sets up camp in the arid Southwest. They're large ants, with the workers hovering around the one-centimeter mark, and thus easy to track with the naked eye. Their colonies, while underground, can be clearly spotted in grasslands because they leave a wide radius of cleared dirt around their nest. The ants get their name from the seeds that they gather, husk, and eat; in extremely dry environments, the ants pull water out of the seeds by metabolizing the fats within. The husks are carried outside the nest and dumped unceremoniously into an ever-growing heap. The ants seem to impregnate this midden pile with the same grease they rub on their bodies, whose scent helps guide foragers back to the nest, like a flag waving high over a fortress.

All of the ants in the colony are female, except for the time of year when the queen produces alates, winged reproductive males and females that fly out of all the different colonies in the area and meet up at some mysteriously divined location. The females mate with multiple males; the males then die (their job is done) and the now-crowned queens fly off to start their own colonies. Each one digs a safe hole in the ground and lays the first round of eggs, feeding the young using her own stored fat reserves; once that vanguard reaches adulthood, they begin to forage and build out the nest, and the queen never emerges into the light again.

You may think of ants as pests, invading your kitchen as easily as they raid your picnic basket, but ants perform a host of valuable ecosystem services. They perform seed dispersal for plants, allowing them to hitch a ride to new territory, and their underground nests made of many interconnected chambers are thought to aerate the soil. They enrich the soil, and in some forests, serve as a defense mechanism for the trees that they live on.

Before they were seen as sugar-hunting home invaders, ants have long impressed humans with their apparent industry and work ethic. "Go to the ant, thou sluggard," the Bible admonishes its readers. In ancient Greek mythology, Zeus turned a colony of ants into a legendary people known as the Myrmidons, loyal and brave warriors who followed Achilles into battle in the *Iliad*. In the deserts of the Southwest, home to the harvester ant, the Native American Hopi people have a story about the ant people, who sheltered their ancestors during an apocalyptic fire reminiscent of the world-cleansing biblical flood. Their forbears, the Ancestral Puebloans, were (rather like the ants) a matrilineal society famous for building homes into the cliffsides—homes that sometimes remind Gordon of the ant nests, perhaps because they were essentially built from the same earth.

But many depictions of ant colonies have taken on a very fascist interpretation—take the animated film *Antz*, where soldiers (mostly male, apparently) bark out orders to platoons of other ants standing in formation. Even Marvel's 2015 movie *Ant-Man*, which paints a more sympathetic portrait of the insects, gives Paul Rudd's character the ability to direct whole swarms of ants with highly complex instructions. This is almost as far from the truth as you can possibly get. Ants do not take orders from on high; they couldn't if they wanted to. An ant's environment isn't regimented, but messy and chaotic. They simply respond to the stimuli in their immediate environment—extremely immediate, because they're generally nearly blind. The insects communicate by touching their antennae to one another's bodies to smell the hydrocarbons coating them. The messages in those molecules, naturally, have to be extremely simple.

Back when Gordon first started studying ants, scientists had a few misconceptions of their own that seem to mirror the pop-culture ones. As in the dystopian novel *Brave New World*, ants were seen as a highly controlled caste society, with its members' characteristics and duties determined at birth (or, in the case of Brave New World's Alphas, Betas, and Epsilons, "decanting"). Certainly that's an alluring idea for humans, given how often in our recorded history we've resorted to castes of various sorts, whether in colonial India or feudal Europe. And it seems to make some kind of sense, given that there can be serious body differences between, say, worker ants and soldier ants, which often have bigger bodies and monstrous jaws for attacking intruders.

That idea is actually made up of two different parts: that an agent's purpose is solely determined at its creation, and that its purpose is singular. It's a concept that long held sway not simply over swarming insects like ants, but over a variety of disciplines, from genetics to neuroscience. A few decades ago, scientists thought that each gene simply coded for a particular characteristic or process in an organism, and a neuron held a particular memory. But as they've learned about the complexities of the molecular chemistry at work, scientists have had to abandon that simplistic idea.

"The idea that each gene just codes for some particular part of the phenotype of the organism has been replaced by the idea, by the discovery, that genes interact with each other and there's a lot of regulation and dynamics and exchanges," Gordon pointed out. "The idea that every ant had its job and did it independently of the others and that every gene is coding for some particular thing—we've had to give up on that kind of thinking throughout biology.

"When I started working on ants, the idea was that each ant was sort of independently programmed to do its task over and over," Gordon said. But Gordon was proposing a radically different idea: that the ants were constantly responding to the actions of other ants around them. "The same kind of transformation in our thinking has happened throughout biology, really accompanied by a revolution in computer science, where people started to think about distributed processes."

Gordon is talking about that concept known as "emergence": the way that large-scale, complex systems can arise out of very simple interactions between many different agents, without any central plan. And in ants, she managed to demonstrate that it was happening using a very simple tool: toothpicks.

In red harvester ant colonies, the ants have a variety of different jobs. Some are patrollers, leaving the nest at the crack of dawn. A patroller serves as the proverbial canary in a coal mine: if they get back safely, that's a good enough sign for the foragers that it's safe to go outside. The foragers head out a little later, following the direction of the patrollers as they search for food sources to bring back to the nest. The nest maintenance workers gather seed husks, dead bodies, and other debris and take it all outside the nest; the midden workers consolidate that refuse in a well-defined heap and imbue it with fragrant grease. Altogether, these outside workers make up about a quarter of the nest population. Others remain inside the nest, performing maintenance and tending to the young. The queen, of course, gives no orders—her job is to serve as the colony's ovaries, using the limited supply of sperm from her one-and-only mating session and keep pumping out more ants.

Now, according to the old way of thinking, foragers were always foragers, patrollers were always patrollers, and midden workers—well, you get the idea. But Gordon decided to test this idea by stressing the capacity of one of these groups, and seeing how the colony as a whole responded. She placed piles of toothpicks near the nest entrance, placement that would be sure to make a maintenance worker grit her mandibles. Sure enough, about half an hour later, they'd all been moved to the edges of the nest's surface. Gordon ran this trial through again—though this time, she painted the bodies of the ants, to be able to identify and follow individuals. Sure enough, when the toothpick pile appeared, requiring a temporary hiring of extra nest-maintenance workers, the numbers of foragers went down. And of course, if she put out food, the number of foragers spiked and the numbers of nest maintenance workers, patrollers, and midden workers went down. And as she followed individually painted ants, Gordon could see that the ants were indeed switching tasks as the situation called for it. Thus, an individual ant's purpose was not set in stone, as the prevailing theory claimed.

But they weren't switching equally between jobs, Gordon realized. There seems to be a hierarchy at work: If more foragers are needed, any of the other workers can switch to the more urgent task. If they need to move a bunch of toothpicks, however, none of the other groups can assist. Instead, more, younger ants come up from inside of the colony to help the nest maintenance workers out. Foragers, patrollers, and midden workers simply hang out until they're needed again. Keep in mind, the ants in these four categories Gordon is describing—midden work, nest maintenance, foraging, and patrolling—make up just 25 percent of the total population of ants. And for the most part, they seem to be exiled to the outer fringes of the nest; they never go deep into the colony. And the ants inside the nest, who are husking seeds and taking care of the brood and maintaining the nest inside—they never seem to leave. Sure, an inside worker might bring some trash up, but they'd generally leave it in one of the outermost chambers, and then one of the outside workers would pick it up from that spot and take it outside to dispose of, like a bucket brigade. And at any given time, roughly a third of the nest's ants are hanging out and doing nothing in a sort of reserve in the mid-levels of the nest, perhaps waiting for instructions.

As it turns out, whether an ant is an innie or an outie seems to determine how they smell. The greasy layer of hydrocarbons coating an ant's body undergoes chemical changes as they spend more time in the sun, and that scent acts as a label for their current job. ("It's like the calluses on a carpenter's hand," Gordon says.) Younger, freshly greased ants stay inside the nest, doing inside chores. As they get older, they gradually migrate up and out as they switch jobs, ultimately spending their old age as foragers, far from the heart of the colony.

Gordon also noticed that younger colonies seemed to be far more reactive during these sorts of disturbances than the older ones, sending a phalanx of ants to deal with sudden messes or windfalls. The older nests, in contrast, seemed to have a far more measured response, able to keep all work flows operating at relatively normal levels.

Perhaps the older colonies' ants were more mature? Hardly. While the queens can live for decades, the workers only live about a year at most. With that kind of employee turnover, there's not a whole lot of institutional memory in a twenty-year-old colony, compared to a three-year-old one.

What did seem to differ was the colony size. In its first year, a colony has about 500 ants; a two-year-old colony, around 2,000 ants. But by the time it hits age five, that number has spiked fivefold, to about 10,000. Some colonies might inch up toward the

12,000 population mark, but for the most part, a colony reaches its "adult" size around age five.

Why would nest size matter so much that it could change the very personality of a colony? It must have something to do with the interactions between the ants—namely, in the number of interactions between them, Gordon realized. Think about a single foraging ant: It wanders around, finds a nice juicy picnic, and returns to the nest with a piece of food. When it reaches the nest, it runs into an ant waiting inside the entrance, which probably "sniffs" the returning ant with its antennae. Remember, the ants have a layer of grease that they spread all over their cuticle-covered bodies while grooming themselves and each other. This smelly grease helps ants identify members of their own colony and distinguish them from those of other neighboring nests—and it also indicates their current task. So the ant hanging out at home smells the returning ant with food. Then another one comes, and another one. At some point, the ant waiting at the entrance gets enough signals from returning foragers that it decides to go.

Gordon tested this theory by taking little aluminum chips, coating them with hydrocarbons and food odor (to mimic seeds brought home by foraging ants) and then dropping them into a nest at what seemed to be the right rate of return. Sure enough, this triggered more ants to leave the nest and search for food. Chips coated in hydrocarbons alone did not have an effect; neither did chips solely covered with food odor. Both cues had to be present. Gordon's experiment revealed just how simple ants are, that they can be fooled by a greasy bit of metal. And it becomes all the more incredible that their systems are so effective, so responsive, so dynamic—so apparently intelligent—even though they are acting with such a minimal amount of information. No matter what the Bible says, ants do not have to be wise to be incredibly successful.

These interactions are key to the older colonies' success. Think of the large colony as downtown in a big city, and the small colony as the main street in a small town. In the city, you're constantly jostled by the throngs of people around you; the rate of interaction is very high. In a sleepy little town, fewer people walk the streets at the same time—and the rate of interactions is much lower. So it's much harder for a stressor to elicit a strong reaction in a large colony, because the ants who might respond are already bombarded with signals from the other ants who are just doing their regular jobs. But in a small colony, with fewer interactions to drown that stressor signal out, the response is much stronger.

"The higher rate of interaction in the larger colony acts to dampen some of the feedback from interactions," Gordon said. "So the outcome is that the older, larger colony is more stable in response."

Whether or not ants are smart in a traditional sense, there's much to be learned from the successful colonies. Gordon comes back to the same study site year after year, a roughly 330 by 440 yard area off a paved road not too far from the New Mexico border. There are roughly three hundred ant colonies in this nearly thirty-acre spot, a number that has remained fairly stable over the years. She's noticed that different colonies seem to have different personalities—personalities that can be easily distinguished in times of hardship. Red harvester ants prefer to work in the early morning hours, retiring back to the nest when midday makes the desert too hot and dry. An insect who stays out runs the risk of dying of dehydration. On the hottest days, many of the colonies cut down on their foraging activity, while others soldier on, gathering the seeds that they need for both food and water.

At first, Gordon assumed that the more active ants must be outcompeting those clearly lazy colonies who couldn't be bothered to bring home much proverbial bacon on a hot day. But she decided to measure their success using a tried-and-true metric:

reproduction. Those colonies that had more daughter colonies, presumably, were the ones that were doing things right. Gordon sampled the colonies at the test site, carefully mapping out the genetic relationships between them. And she found that the overly active colonies were not doing as well as she'd thought. In fact, the queens whose workers were more reluctant to forage on toasty days ended up being mothers, grandmothers—even great-grandmothers. In a place like Arizona, playing it safe pays off, because an ant foraging for seeds on an excessively hot day will actually lose more water from its body through evaporation than it can gain from the seeds it gathers. It's the simple kind of evolutionary math that becomes hard to argue with.

Gordon knew that what she was looking at was the passing of extremely basic bits of information driving a larger system. She already knew that there were a lot of potential parallels that could be drawn between the way ants behave in relation to the colony and the way cells behave in an embryo—and how neurons behave in a brain. Ants communicate by exchanging chemical signals with their nearest neighbors; so do neurons. A foraging ant inside the nest won't go searching until she smells a certain number of foragers returning. Similarly, a neuron in the brain may not fire until it's received a certain number of stimuli.

Think also about the colonies that foraged even on the hottest days—the Type A colonies, if you will—and those that chose to stay indoors, the Type B colonies. Perhaps that difference in personality is defined in large part by a difference in individual ants' sensitivity to stimulation—in the case of the foragers, the number of encounters an ant must have before deciding to go out foraging herself. If that number is high, an ant will be less likely to be moved. If it's low, she'll hop to it at the drop of a hat. (I also see a similar parallel in the behavior of neurons; perhaps in different people, the same type of neuron might take more or less stimulation before it fires, and collectively this might result in some significantly different behaviors and personalities between two individuals.)

In any case, both of these complex emergent systems are the result of sending and responding to simple bits of information—tiny packets of data. Neurons and ants are living networks; and Gordon had come up with the algorithm that described the foraging aspect of the ant network. Perhaps, she realized, this algorithm could be useful for other types of networks—even nonliving ones. So she turned to Stanford computer scientist Balaji Prabhakar for insight.

Prabhakar was bemused at first. "You know how it is, it's like, when someone from biology wants to meet you . . . you meet up out of curiosity partly, and partly out of politeness, not expecting much to happen," he told me.

But after mulling it over for a day, he realized that there really was a similar man-made system: the Transmission Control Protocol, or TCP. Together with the Internet Protocol (IP), it serves as the basic communication language of the Internet, and might be familiar to you as TCP/IP. TCP helps to regulate data congestion on the Internet—and, it turns out, operates in much the same way that the ants do.

Here's basically how it works: Say Ben is trying to send a large package to his friend, Jerry. Now, to send this hefty parcel along, Ben chops it into tiny manageable pieces and labels each one with a number and sends them off, one by one. When Jerry gets each packet, he quickly dashes off a little thank-you note—an acknowledgment that the package was received. Ben, because he's a little insecure, only sends out the next packet after he gets that acknowledgment that the last packet made it through. This sets up a really effective feedback loop that actually causes the network to self-regulate on this local scale. After all, if Jerry doesn't send that acknowledgment back in a timely matter, it could mean any number of things—that the postal worker never made it to Jerry's house, or that he had trouble carrying the thank-you card back to Ben. Regardless of the explanation, it means that the mail carrier's ability to carry messages back and forth—the network's bandwidth, in short—is too low for Ben to send that next packet along.

TCP was essential in allowing the Internet to grow from a small number of computers hooked up between a few researchers to the global network it is today, because it's scalable in a way that a centralized network is not. Think about it: Under a centralized system, the larger the network, the more connections need to be monitored and maintained, which means way more work for the central hub, where all that constant computation would have to be done. But in a distributed, headless network, as in TCP, each node trying to send information to another node is only worrying about its local environment—the connections it's immediately coming into contact with. That's true no matter how small the network is, or how large it eventually becomes.

Now, this system should sound very familiar to you at this point. It's very much akin to what the harvester ants are doing when they're waiting at the entry of the nest, touching their antennas to those of returning food-laden foragers, and deciding whether or not to go out. In both cases, the rate of return is what tells the ants and Ben whether or not to keep sending more. If computers have the Internet, ants have what Gordon called the Anternet.

Unlike the Internet, which has really only been around for a few decades, the Anternet has probably been around for roughly 150 million years. Perhaps if humans had been smart enough to look to the ants a few decades earlier, they'd have saved themselves many years of research and development.

Gordon learned all of this from studying a single ant species. What if there are other clever networking algorithms used by different species that haven't been devised by humans, but have been used by ants for eons? Harvester ants' simple networking rules were the result of millions of years of evolution, responding to the environmental pressures at hand: climate (specifically, the operating costs it imposes), food resources, predation, and competition for those food resources. Ease up on some of those pressures and push down on others, and you might end up with a very different algorithm—one that might prove useful for us, too.

"There are fourteen thousand species of ants, of which we've studied about fifty in detail," Gordon said. "And I think there are a rich wealth of algorithms to be discovered that we could use for how they search collectively, for how they regulate activity, for how they create networks—all kinds of amazing things that ants do."

The red harvester ants are the species she knows longest—and she's still learning new lessons from them. When she first started studying these ants, no one knew how long colonies lived. After two decades studying them, she concluded that the queens must live fifteen or so years. A decade later, those colonies were still going strong. But over time, Gordon has branched out to study a wide range of species in different environments—and she's found that their collective intelligence shows different thought processes, if you will, depending on the ecosystem's particular pressures.

In the desert, the ants are dealing with scarce resources, and a manageable amount of competition. This means that they tend to stay put and save their energy unless a positive stimulus comes along. But for ants that live in lush forests, where the weather is comfortable but there are many other ant species competing for the same resources, the inverse is true: ant colonies tend to keep going unless they encounter a negative stimulus.

That's very much the case for the arboreal turtle ants in the tropical forests of Mexico, another species Gordon visits during her summers off from teaching. In these humid climes, the ants don't need to worry about getting baked as they search for food, and in the forest canopy, there seems to be plenty in easy reach. But there are a lot of other ant colonies, including many different species, going for those same resources. Gordon studies the turtle ant *Cephalotes goniodontus,* known for its distinctive headgear: a broad shieldlike helmet on top of its face. These ants forage among the trees, rarely coming down to the forest floor. She and her assistants had to bring a ladder to try and examine the nests and trails, and even then some of them remained far out of sight, among the treetops. These ants have wider-ranging appetites than their seed-harvesting cousins. They gather all kinds of edible treasures, including lizard and bird droppings, caterpillar frass, bits of fungus, and lichen. Gordon saw them foraging for nectar and other plant fluids, which they'll sometimes stop and share with their fellow colonials that they meet in the middle of a trail. As for what kinds of bait they'd take, protein like egg or fish never worked, although sometimes cake did. They seemed to take to cotton soaked in sweet hibiscus juice—although the most tempting bait appears to have been human urine (and no, I'm not sure whose).

Unlike the single nest of harvester ants, turtle ant colonies actually have multiple nests in the cavities of trees that often seem to have been previously dug by another insect, and they use tiny holes in the branches as doorways to these nests. This is what the headgear is for: If another ant species, following trails left by the turtle ants, comes across an opening, a helmeted turtle ant will immediately shove its body into the doorway and plug the entrance using its broad, flat head. Gordon thinks that their behavior could be a model for network security, envisioning a system that's dynamic and responsive to threats as they appear rather than trying to build an impervious, static monolith of protection—an impossible task, in any case.

Armored as their soldiers may appear, turtle ants aren't particularly warlike; in fact, they're peaceful almost to a fault. If another species of ant follows the turtle ant trail to the food source, they'll abandon that trail until the other species goes away. That's almost surprising, given that turtle ants do a whole lot of legwork to build their trails, moving from a twig on one tree to a vine it happens to be touching and then following the curve to the leaf of another tree, where a food source may lie. These trails are convoluted, often two to five times the straight-shot distance. In one colony, one trail whose linear length was 39.4 feet had an actual length of 162.4 feet, which included some twenty-nine transitions from one bit of vegetation to another.

These trails aren't just long—they're also fragile. A leaf can get eaten, a twig can break, the wind can blow a vine away. The ants seem pretty unfazed by this, as Gordon documented in a study of the species over several rainy seasons from 2007 to 2011. She described one particular trail that relied on a broken branch held in place by some tangled vines:

> When the wind blew the branch away, the connection was lost. Ants that arrived at the tree when the branch had been blown out of place waited at the gap, like passengers waiting for a ferry, until the wind died down and the branch came back, and then stepped onto it.

There are times, though, when no amount of patience will cure a broken trail. The ants must then find and build a new trail to the food source—a task they're remarkably good at, considering that they've got terrible vision and are living in a pretty disordered environment. Gordon thinks that learning the simple rules by which they make new trails and "repair" old ones could lead to an algorithm that could help humans build more resilient networks of all kinds, able to withstand these kinds of sudden and unpredictable breaks. It could also help shed light on how similar processes occur in the brain, she added.

"When there's a break they quite quickly manage to repair that and actually over time prune it so that they get a quicker path," she explained. "And that pruning process looks like synaptic pruning in the brains of children. It also would be a very effective way to design any kind of network where information is passing along and there might be a break—so systems of cables, communicating devices."

The biologist has even sent ants into space, in which astronauts aboard the International Space Station monitored how the pavement ant (*Tetramorium caespitum*) changes its search pattern in zero gravity. Inspired by that experiment, Gordon also designed an extremely simple "habitat," just like the ones used on the space station, except made of everyday materials: paper, foam board, and Plexiglas. The entire habitat is held together with binder clips. Students around the world can use these habitats to document the collective search behavior of whatever species are native to their town, and (if they're so inclined) send the results back to Stanford researchers. Perhaps, with the help of students around the world, they'll be able to crowdsource the data that scientists can use to discover different search algorithms used by species in a variety of ecosystems.

"The more I study ants, the less I think we know. But . . . we are certainly becoming more open to thinking about it in new ways," Gordon says. "So I don't think it's quite a matter of amount of . . . stuff known, but instead the ability to think in ways that make it possible to see what's happening better."

Gordon, at this point, has studied many different species, so I ask her which is her favorite. This question, of all the ones I've asked, seems to stump her.

"You know, I've spent a lot of time with harvester ants—I know them well, I've grown up with them," she says after a moment. "The turtle ants that I study actually are kind of amazing, they're such weird-looking ants . . . they have such an amazing combination of appearing to react quite slowly but very rapidly solving problems.

"There's some species I have not been able to like, like the Argentine ants," she quickly adds. "They're really interesting species but I can't really like them."

"Why not?" I ask.

"Why? Because they're so—imperialist," she says, with a little laugh. "They're very persistent, and I sort of have to admire that. But . . . they're [an] invasive species and they wipe out the native species, so when they're around, the other species aren't there and I feel like they've taken more than their share. They're greedy."

Argentine ants and fire ants are both on Gordon's hit list, but she still studies them. After all, the invasive species have been remarkably successful; there must be something they're doing "right," for lack of a better word. Gordon grew up in Miami Beach, Florida, right around the time that fire ants were beginning to really build upon the foothold they'd established in the United States several decades before in Mobile, Alabama. In the decades that followed, the marauding ants spread through at least fourteen states and Puerto Rico, moving from the Southeast through the Southwest and all the way to California. They're an aggressive, stinging species; the welts their bites leave will last for days;

and until recently, their name *Solenopsis invictus* (which means "unbeatable") has seemed well earned.

They are, in a lot of ways, an academically fascinating species: David Hu of Georgia Tech has studied how, in their native South America, these insects can grab onto each other, mandibles hooking onto one another's legs, and turn themselves into giant balls that can serve as a life raft during treacherous Amazonian flash floods. There even seems to be an internal structure of sorts, though a ruthless one—the workers put their babies at the bottom of the life-raft, perhaps because they're more buoyant and help keep the entire raft from sinking. If a few of them drown in the process, so be it. So long as the queen is safe, more can be made. Hu is interested in the simple rules that govern the way they form rafts, bridges, and other structures—how these rules could help engineers design smarter, more dynamic materials.

Hu is a mechanical engineer and a little more dispassionate about ants than your typical myrmecologist; Gordon remembers one time he came to a social-insects conference in Australia and showed the gathered scientists a video of a mass of ants being (very gently and non-lethally) squished to demonstrate their material-like properties—causing the gathered biologists, who are typically quite fond of the insects, to cringe. "It was just funny to see," she said of the contrast.

Gordon herself is interested in imperial ants like the fire ant and Argentine ant because of the search patterns they use—and how they change depending on ant density. For example, when fire ants are densely packed, they take very winding paths, thoroughly searching small areas. But when there are relatively few ants from the same colony in a large area, the ants walk in long, straight lines, choosing to cover a lot of ground rather than explore it thoroughly. How do they modulate their behavior? You guessed it: by the rate at which they run into their sisters and touch antennae.

Even from these invasive species, then, Gordon thinks valuable lessons can be learned. She also thinks we shouldn't give them too much credit. Humans can probably shoulder much of the blame for the rise of fire ants, according to *The Fire Ant Wars*, a book that chronicled the political and philosophical conflicts surrounding the spread of the invasive insect. Mississippi congressman Jamie Whitten, who chaired a committee that controlled the U.S. Department of Agriculture, was one of those who fought to keep using pesticides during the 1960s and 1970s, even as environmentalists increasingly warned that the chemicals were highly toxic to other animals. (Whitten appears to have had a very cozy relationship with pesticide manufacturers; a book that he apparently wrote as a rebuttal to the environmental classic *Silent Spring* turned out to have been "conceived and subsidized by pesticide industry officials," according to the *New York Times*.) For example, the pesticide Mirex, like previous chemical agents unleashed upon the fire ant, proved to be very effective at killing things—just not the right things. After it was applied, dead birds and mammals started showing up on the sides of roads. Lab results from the Environmental Protection Agency found Mirex in the fatty tissues of more than a third of humans tested in the Southeast.

"It was really carcinogenic stuff," Gordon said.

Ironically, such pesticides helped clear the way for the fire ants to take over, because the chemicals killed native ants, leaving their territory free for the taking. Without this helpful human hand, it's possible that the competition from all the other species in the area would have kept the fire ants more or less in check. When fire ants spread through Orange County in California in the 1990s, there was yet another discussion, under Gov. Gray Davis, about whether to spray another pesticide. Apparently the lessons from the Southeast had not stuck.

Right now, the Golden State is dealing with the ant version of *Alien vs. Predator.*

"Here in California the fire ants are up against the Argentine ants," Gordon explained. "And it's not clear what's going to happen. The fire ants need more water. . . . So one vision is that every golf course you know along the California coast will be filled with fire ants but the Argentine ants will be all around it."

But, in a silver lining to the state's water scarcity problem, both species have suffered in the drought. And Gordon has begun to find native species who are mounting a resistance against the invaders—including a Northern California native, the winter ant, that can squeeze droplets of poison onto offending Argentine ants with a brutal 79 percent kill rate. She's starting to see hope in other parts of the country, too.

"One of the interesting things that's happened in the Southeast is that the native species have actually pushed back," she said. "There are places where the fire ants were, and are no longer."

One of the researchers who pioneered the notion of ant-based algorithms was Marco Dorigo, an Italian computer scientist and roboticist at the Université Libre de Bruxelles. Back in the late 1980s, when he was a graduate student at Politecnico di Milano, the idea of genetic algorithms were becoming a topic of great interest among computer scientists. This in itself was a bioinspired idea, and one that began to play with the idea of emergence. Genetic algorithms sought to take advantage of the Darwinian process of natural selection, which drives evolution, to build more effective computer code. Computer programs can be found everywhere in our lives today—in the supercomputers that scientists use to develop maps of the universe, in the laptops we use, in the cars we drive, and the refrigerators in which we keep our milk. Every single one of the programs in these objects is vulnerable, either to hacking or to human error.

The problem is, it's difficult for any one person, or even a team of people, to deal with this increasing complexity—and that's a lesson that applies to a whole range of fields, not just computer science. In the middle of the twentieth century, in the wake of the splitting of the atom and the discovery of DNA, many scientists argued that much of the fundamental science in such fields as physics had already been done, and all that remained was finding novel ways to apply those basic lessons. Aside from the erroneous assumption that the need for basic science has passed, this idea also assumes that if you know those basic rules, then you know how things operate at every intermediate stage and scale above that—as if, having learned the limited set of characters in an alphabet, you might be able to construct any word, any sentence, any paragraph, any book you want. Conversely, any complex system should be easily broken down, reduced to its component parts.

In a famous 1972 article in *Science* entitled "More Is Different," physicist and Nobel laureate Philip Warren Anderson took that argument apart bit by bit.

"The constructionist hypothesis breaks down when confronted with the twin difficulties of scale and complexity," Anderson wrote. "The behavior of large and complex aggregates of elementary particles, it turns out, is not to be understood in terms of a simple extrapolation of the properties of a few particles. Instead, at each level of complexity entire new properties appear, and the understanding of the new behaviors requires research which I think is as fundamental in its nature as any other." These ideas began to take flight in the decades that followed. John Holland, a computer scientist and a giant in the study of complex systems, began to think about ways that the lessons of complexity from nature could be applied to computer programming. He also became interested in the ideas of John von Neumman and Stanislaw Ulam, both scientists who worked on the Manhattan Project and who while at Los Alamos National Laboratory came up with the idea in the 1940s of what were called self-reproducing automata: computer code that held the instructions to make a copy of itself—offspring, in short.

"In other words, von Neumman had proven that something like DNA must exist—and had done so before Crick and Watson discovered DNA's structure," Scott Page, a scientist at the University of Michigan, wrote in a remembrance of Holland, who died in August 2015.

Holland's 1975 book, *Adaptation in Natural and Artificial Systems,* would go on to influence thinking in a wide range of fields, from neurobiology to computer science. Among the concepts he laid out were genetic algorithms, an artificial analogue to the mechanisms powering natural selection. By way of example, let's say you have a species of duck that needs to eat algae or seeds, but because of a recent blight the only species around is kind of poisonous and the remaining plant seeds require a specially shaped beak. In this species of duck, some may have better shaped beaks; some may be able to withstand the poison; some may have neither genetic advantage. But as a rule, the ones that can't take the poison and have the wrong kind of beak will die before they can produce. The rest of the ducks will mate, mixing their genes in the next generation of ducklings. Those with the right combination of poison-resistance and beak shape will be more likely to survive and reproduce, and so on. Along the way, random mutations may occur as the ducks swap genes, which may disappear if they're useless but survive to be copied if they offer some other useful advantage.

Holland adapted this into computer programming. The result: A system where different variations on a bit of computer code could run, then be evaluated based on their success, and be given a probability to "reproduce" based on that success. The most successful program might have a 60 percent rate, the runner-up might have a 30 percent rate, and so on. The worst performer was typically knocked out entirely. The remaining programs would "mate," making new programs by swapping bits of their code at a randomly chosen interval. On occasion, a mutation might be introduced into the resulting "offspring" code during the copying process, just as it might happen in a cell. They would keep doing this until the programs had "evolved" to a point where they could most effectively perform the desired task. It's an idea that is ubiquitous in computer science and robotics today: rather than trying to meticulously design the perfect code, scientists instead let "evolution" do the work.

Back in the 1980s, Dorigo was interested in this idea because of the reinforcement mechanism—the way that you could "reward" good behavior in a program (in this case by giving it a higher probability of reproduction) and so create a positive feedback loop that drives a system to the desired result.

"At that time," Dorigo said, "the field of genetic algorithms, of evolutionary computation, was blossoming. So there was a scientific environment in which there were already starting to be a number of researchers who were taking inspiration from nature to solve real-world problems."

When he was first starting his Ph.D. around 1989, Dorigo went to a talk by biologists that showed him another reinforcement mechanism: the pheromones that ants leave along their trails to guide one another.

"These biologists were explaining how ants are capable of finding the shortest path between a food source and their nest, just using pheromones, without having any knowledge of a map of the environment," Dorigo recalled. "And it was at that time that

I had the idea that maybe a similar mechanism could be used to solve difficult mathematical problems."

One such sticky problem was the Traveling Salesman problem. Here's how it goes: A salesman needs to visit X number of cities on his route as he sells ties, and he wants to pick the shortest possible route between them, without any repetitions, and end up at his starting point. What route does he take?

It actually takes a ridiculous amount of computing power to find the best answer to this question. And because it's a combinatorial problem, it means that adding even one city to the list can increase the computing time exponentially. If you were able to solve a problem with fifty cities with a one-day computation, he said, adding one more city to the list could hike the computation time to two years.

"Fifty-two can take [more than] one thousand years," he said. "There is this exponential growth in the complexity of the problem, which makes it, basically, practically unfeasible to find the optimal solution."

But ants are remarkably good at finding the shortest path, which they can do thanks to pheromone trails. Here's the basic idea: There are two trails to a food source, one short and one long. An ant, moving at random, may take the long trail, and drop pheromone behind her as she comes back with food. This means more ants will follow the smell-trail on the longer path, which means extra legwork for everyone. But another ant might randomly wander off and find the shorter route, and bring the food back faster. More ants will then follow. And since the ants on the shorter path will be bringing the food home more frequently, they'll end up laying down more pheromone overall, which means their trail will soon smell stronger than the longer trail. Keep in mind, pheromone molecules eventually evaporate, so if the trail isn't being refreshed by ants bringing home the proverbial bacon, it will fade away.

Dorigo designed an algorithm that used imaginary ants that would crawl along a point-and-line "map" of the cities in the Traveling Salesman problem, with a few modifications. If an ant found a route it really liked, it would drop a lot of pheromone. If it was on a route that was kind of long, it would drop just a little pheromone. This pheromone could be programmed to evaporate, much more quickly than it would in nature. With many simple ants doing this in parallel, the way they might while foraging around the nest, they would quickly converge on a good solution. The route they chose might not be the absolute shortest possible—but it would be good enough to work with, Dorigo said. That, too, is akin to real life; ants don't necessarily find the absolute best foraging solution, but they do find one that is better than most.

"If you are a traveling salesman and you want to visit one hundred cities, and you know that you have to wait for your computer to provide the optimal answer in one thousand years, you don't care about the optimal answer, it's not useful," Dorigo pointed out. "But if the computer can provide a reasonably good answer in a few minutes, then the solution is good enough, and it's useful. Because a few minutes, you can wait."

Thus, ant-colony optimization was born. Since then, Dorigo and other researchers have used algorithms inspired by different aspects of swarm behavior, from foraging to brood sorting, to solve a range of real-world problems. For a company working with Italian pasta maker Barilla, they designed routing software that tells them which trucks to use, where to send them, which roads to use, and what types of goods to put on the trucks. It's also been used by phone companies to route phone calls when the lines get congested or there's a breakdown in one of the routes. Recently, scientists working with Unilever reported an ant-inspired system to schedule different tasks in a large plant.

Dorigo's work has been cited around eighty thousand times,

and he estimates that roughly two-thirds of that is for his work on ant-colony optimization. But lately he's turned to robotics—designing and building swarms of insectlike bots that, in spite of their limited sensing capabilities, can complete a specific task (such as "move this large object") without needing an overseer. It's not easy work—not only are you dealing with computer programming, but you also have to deal with the vagaries of mechanical failure—and it pushes the development time on a project from a few months to a few years. But Dorigo expects that in the future, such robots will be cheap and effective tools for exploring oceans, exploring outer space, constructing buildings, performing rescue missions, or perhaps just keeping your home clean. For the moment, however, that work is in very early stages.

Either way, he said, it's important to keep turning back to nature for inspiration, because you can't apply lessons in collective intelligence without learning those lessons in the first place. Collaborations between computer scientists and biologists are key—but those collaborations need to be wisely chosen, he added.

"You cannot take any engineer and let him work with any biologist," Dorigo said. The biologists that feel "closer to engineering and engineers that feel close to biology . . . you have to put these people together."

Dorigo worked with one such biologist, Guy Theraulaz, who came of academic age around the same time Dorigo did. Like Deborah Gordon, Theraulaz was looking to study cellular structures—neurons in the brain, in his case—in the 1980s, a time when there were few tools to do so. But luckily for Theraulaz, who was getting his master's degree at the University of Provence in Marseilles at the time, there was a group in the same institute that was studying animals that allowed him to explore many of the same questions: wasps.

Wasps, which are distant cousins of ants, are also fairly dumb as individuals. But an individual neuron is pretty stupid, too. It's only through their collective interactions that we get something from the entire organ of the brain that we call "intelligence." Studying these insects, relaying signals only between their nearest neighbors, would allow him to probe the secrets of these complex systems.

It strikes me, briefly, at how ironic it is that a lack of technology is what spurred research into collective intelligence. If scientists had had the sophisticated instruments and imaging techniques they have today that allow them to document the behavior of a cell in action, would we know as much about swarm smarts as we do today?

None of this is to say that studying wasp behavior at that time was easy, even for small colonies of twenty or so insects, housed in glass at the lab. They had to spend eight hours a day carefully watching each insect's movements and entering them into a very rudimentary computer. Theraulaz himself had to develop the software that would allow them to track the bugs.

"That was a very exciting period, but you have to imagine, at that time, we didn't have a Macintosh . . . that was the beginning of [personal] computers," Theraulaz said.

Theraulaz went on to pen a book on swarm intelligence with Dorigo and their colleague, Eric Bonabeau. And he also worked on swarm robotics systems. But in recent years he's spent a lot of time studying less overtly swarmlike animals, including the mathematical rules governing the movement of herds of sheep and schools of fish. He's become especially interested in how collective intelligence can emerge in human populations, and has been studying the behavior of large groups of students in France and Japan—results that sadly, he can't tell me much about yet, as they've yet to be published.

Ants (and bees, and wasps) move randomly through the

world, and they have to deal with random perturbations in the environment that may disrupt a stable system. They rely on both positive and negative feedback, from the environment and from each other, to prune out most errors in judgment. They don't even need to communicate directly with each other to do this—an ant can leave pheromone on the ground now and another ant can pick up the signal later. As described in the previous chapter, a termite may drop a piece of soil on the ground and another termite coming by later might then drop its mouthful of soil on the same spot, responding to an internal rule that says "drop soil when you see a soil pile." It's a concept called "stigmergy," and it's key to successful communication in these swarms.

In human groups, however, communication often seems to go horribly awry. In spite of being the only species known to have a full-blown language system, we seem to allow bad information to propagate just as quickly and widely as good information. A single bad review, no matter how libelous, might ruin a fledgling restaurant's reputation on Yelp. Pictures of historical figures like Harriet Tubman or Abraham Lincoln are tagged with a fake quote and quickly saturate social media, with no way to stop the flood. And this isn't simply a product of the Internet: since the emergence of language, rumors have spread through populations like a virus (unstoppable by something so weak as actual facts) until they run their inevitable course.

Here's the strange thing about human groups. If you crowdsource something like an estimate—"how many beans are in this jar?"—and you take everyone's guesses, no matter how wild, they will average out to a remarkably accurate judgment. But if you allow all the guessers to talk to one another, Theraulaz said, that final estimate can end up being wildly off.

Humans don't seem to have a negative feedback control mechanism to keep bad information from propagating, the way that ants do, in that the pheromone evaporates. For ants, bad information does not live forever. (Neither does good information, but it will presumably keep getting refreshed as more ants with the right information keep dropping pheromone on their paths.) In some cases—perhaps, for online megastores like Amazon or business review platforms like Yelp, anywhere that crowdsourced feedback is used—this could potentially be addressed if researchers developed algorithms for websites that could treat reviews like informational pheromone, evaporating after a specified time period.

There's also another reason that human populations are not so good at regulating communication, he pointed out: They've become too big to be able to do it. Look at what happens to ants that live in tiny populations, compared with the species that live in the insect version of megacities. In ants that live in nests with less than a dozen individuals, the individual ants often have highly developed cognitive abilities—big eyes, big brains, and they often have a rich repertoire of signals they can use to communicate with each other. But if you look at the species that live in enormous colonies, where one nest can hold twenty million insects, the individual ants' cognitive abilities are severely reduced. They're often completely blind; they rely on a very small number of chemical signals and they exchange extremely limited, simple bits of data.

"They filter the information, and they develop the right answers to a very small repertoire of small signals," Theraulaz explained. "That's the way they cope with the scalability problem. And this is what we want to do now with humans. Because in humans, we are in a period of our evolution in which we interact more and more. We have created the Internet; we have created, now, portable computers, we have created smartphones. We are constantly interacting with each other more and more, but the problem is that we do not filter the information."

It's such a counterintuitive idea—could we simply have too much information to make good decisions? Certainly, I know I feel overwhelmed when reading a dense menu stuffed with more options than I could ever need, and I'm grateful when the choices are limited and my decision is quick. Still, in some ways it seems antithetical to living in a democratic society, which relies on the free flow of information to work.

And yet, there are already examples that seem to indicate that less information is not necessarily a bad thing. In 2014, a National Public Radio reporter visited a pharmacist living in suburban Maryland named Elaine Rich who, at the time of the reporter's visit, just happened to be estimating the flow of refugees in Syria. (Among the other questions she considered: would North Korea launch a new multistage missile before May 10, 2014?)

Rich and about three thousand volunteers were part of the Good Judgment Project, an initiative started by three psychologists along with experts in the intelligence community to tap into the so-called wisdom of crowds. The predictions made by the volunteers often exceed the accuracy of trained CIA analysts—even though, unlike those trained experts, they don't have training, experience, or the significant amounts of classified information available to those agents. In fact, the pharmacist said she didn't even go around looking for obscure sources on the Internet; a simple Google search was all she needed. When it came to making good guesses, less was more. And Rich's guesses were really, really good—her estimates places her in the top 1 percent of the three thousand forecasters.

The psychologists cautioned that, in spite of the project's success, it wasn't clear whether this will work in all situations. Still, it offers a hopeful sign that navigating this complex, messy world might not be as complicated as we think.

PART IV

SUSTAINABILITY

7

THE ARTIFICIAL LEAF

Searching for a Clean Fuel to Power Our World

Dropping one mile down a shaft into an underground mine introduces you to a totally new shade of darkness. On the surface of the Earth, even at night, you always know that the sun is right around the horizon. Buried beneath hard-packed earth, that certainty flickers and winks out.

Hidden deep in this former gold mine in South Dakota was a tank filled with super-purified xenon, which scientists would be using to hunt for dark matter. Here, the thick layers of dirt and rock could shield the detector from the constant radiation bombarding the Earth's surface. The interactions the physicists were searching for would be extremely faint, and it was of utmost importance that no outside "noise" drown out those readings.

The dim tunnels were intermittently lit by harsh fluorescent light. The exposed rock was covered with fine, choking dust. It was hot, and the air was stale—completely inhospitable to life. And yet, as we followed the iron rails that ran across the tunnel floors, I noticed a little flash of green in a patch of exposed, dry dirt: a tiny little plantlet, around two inches high, with nothing more than the two pale round leaves announcing that it had just recently emerged from its seed.

This little plant stopped everyone—the veteran mining workers who helped lead us through the tunnels, the scientists who spent their days in the white, plaster-coated laboratory caverns. In my bulky mining suit and heavy boots, helmet askew, I still managed to get on my knees and elbows and photograph this little miracle. (While I was snapping away, one of the experimental physicists took a picture of my rear end up in the air and later sent it to me with the caption, "for posterity's sake.")

His tongue-in-cheek caption was prescient: long after the newspaper story ran, this little moment has stuck with me, because it's a testament to the incredible, tenacious power of plant life—these living things that spin sugar out of light and air.

Plants are autotrophic life forms—they can make their own food. Everything else on the planet that isn't so self-sufficient has to eat another living thing to survive. Some animals eat plants, and then other animals eat those animals, and when they die, other animals eat their dead bodies, whose nutrients are ultimately returned to the soil. It's a virtuous cycle of growth and death, but without autotrophs like plants constantly injecting new fuel into the system, it would quickly collapse. Whether you're eating a salad or a steak, you're ultimately eating the products of sunlight.

Believe it or not, terrestrial plants are a relatively recent addition to the tree of life. While simple photosynthetic microbes that resembled today's blue-green algae probably emerged in the water around 3.5 billion years ago, plants only made it to land around 450 million years ago. I find this shocking, given that plants as we know them make up the backbone of virtually all terrestrial ecosystems on earth. Photosynthetic organisms, in the water and on land, fundamentally transformed the atmosphere, pulling out greenhouse gases and pumping in oxygen, ultimately giving humans, and all other oxygen-breathing creatures, enough air to breathe.

Plants have changed more than our atmosphere, according to a February 2012 piece by the editors of the journal *Nature Geoscience*. Plants have actively shaped the geochemistry of our soils and our oceans, breaking down minerals locked in rock and releasing fresh nutrients into the environment. They've also sculpted landscapes: before plants took root on terra firma, rivers were simply broad sheets with poorly defined boundaries; plants' root systems controlled the edges of rivers, giving them definition, and sediments from the organic matter they released changed the shape of river deltas.

"Without the workings of life, the Earth would not be the planet it is today," they write. "Even if there are a number of planets that could support tectonics, running water and the chemical cycles that are essential for life as we know it, it seems unlikely that any of them would look like Earth."

Their conclusion: even though astronomers keep searching the skies for an exoplanet that could be an Earth-twin, good luck—we'll never quite find one.

"As we delve deeper into the Earth's past," they wrote, "it becomes clear that many of its apparently original environmental features appeared relatively late, and were brought about by the evolution of life."

So as far as we know—and as far as we may ever know—there is only one Earth. And we're stuck on it for a long time, indefinitely. But we're fast in danger of overrunning it. A thousand years ago, humans could have looked up into the night sky and seen a dazzling array of thousands of stars, some of which could have hosted another world with life. Today, the sky seems sparsely populated; many of those twinkling objects are actually satellites orbiting Earth. The stars haven't died—they're hidden behind the air and light pollution our civilization has produced.

The most dangerous pollution, however, might be the stuff

that starlight passes through with ease. Gases like carbon dioxide, which are transparent to visible light, are responsible for what's known as the greenhouse effect. Sunlight passes through these gases, hits the surface of the earth, and is absorbed by the ground and reemitted as infrared (or thermal) radiation. But instead of escaping from the ground into space, that thermal energy gets caught by the carbon dioxide molecules in the air, which sends it right back down to earth. As a result, the global temperature has risen sharply in the last century. The iconic hockey-stick graph says it all: nearly a millennium of slowly declining temperatures, and then, just a century ago, a sudden turn straight up.

That dramatic rise in temperature is melting the polar ice caps. It's causing sea levels to rise. Florida, whose governor Rick Scott reportedly banned his employees from using the term "climate change," may see much of its coastal property sink beneath the sea in the coming centuries. An estimated one-fourth of species on earth are at risk of extinction if current warming trends continue through 2050.

When we burn fossil fuels—oil, natural gas, coal, any concentrated flammable energy source made from long-dead living things—we're undoing millions of years' worth of carbon storage. That newly released carbon dioxide doesn't just accelerate the greenhouse effect; it's also absorbed into the oceans, causing them to grow acidic, killing off the species that need calcium carbonate minerals to build their skeletons and shells.

Consider this: The worst extinction event in the history of life on earth was not brought on by the violent impact of the asteroid that killed the dinosaurs roughly 66 million years ago. It was the inexorable acidification of the world's oceans, triggered by the massive release of carbon from Siberian Trap volcanism some 252 million years ago. As carbon dioxide seeped into the world's oceans over a period of 60,000 years—with a particularly bad spike in the last 10,000 years of the event—it ended up killing more than 90 percent of all marine species (and more than two-thirds of land animals), wiping out entire lineages whose existence we only know about because of the fossils they left behind.

Luckily, the supplies of conventional fossil fuels are limited, to some extent—this terrible die-off, known as the Permo-Triassic Boundary mass extinction, probably required on the order of 24 quadrillion kilograms of carbon, and researchers estimate there's only about 5 quadrillion kilograms of carbon available in fossil fuel form today. But the problem during the P-T Boundary event wasn't just the amount of carbon being injected into the oceans—it was the speed with which it was happening, particularly during those last ten thousand years. The swiftly changing chemistry of the oceans left species with very little time to adapt—with deadly results. Today, human-generated emissions are causing ocean acidity levels to rise at roughly the same rate as during that extinction event.

So coming up with a viable renewable fuel—one that doesn't release more carbon into the air—is not just some theoretically interesting question. It is vital to the survival of the planet we inhabit.

The desire to learn the secrets of plants to create a clean, renewable energy source goes back more than a century—long before the oil crises and political theater that characterize its buying, selling, and controlling on today's global market.

In 1912, an Armenian Italian scientist named Giacomo Ciamician put forth the idea of solar fuels in a prophetic speech at the VIII International Congress of Applied Chemistry held in New York. In the address, published in the prestigious journal *Science*, Ciamician suggested a future filled with thrilling possibilities and issued warnings that were chilling in their foresight.

At the time, the human population on earth sat at roughly 1.7 billion—less than a quarter of its current 7.1 billion. The industrial revolution, revving its proverbial engine in the nineteenth century, was transforming Western nations into economic powerhouses, and Africa, India, and China still seemed full of raw material (or human bodies) for them to exploit. Coal, which had begun fueling the industrial revolution in earnest in the eighteenth century, had replaced wood as the fuel of choice. Our understanding of the toxic environmental impact of industrialization—rising global temperatures, carbon dioxide emissions, and ocean acidification—were several decades and at least two world wars into the future.

But the soot billowing up from cities was already leaving its mark on Europe (and, though they didn't know it at the time, darkening and melting mountain glaciers in the Alps, decades before climate change became a grim reality). Coal was the "fossil fuel" of Ciamician's lifetime, powering the vast growth in industry and transportation.

Even then, Ciamician saw that this fuel, which produced so much filth, was a limited resource at best. The more of it that was dug out of the ground, the higher the prices would rise, because miners would have to dig ever deeper for fewer and fewer reserves. The rapacious rate of progress did not escape the chemist. If you read his words and replace "coal" with "fossil fuels" (to include oil and gas), it sounds uncannily like the dire warnings issued by renewable energy researchers and environmental scientists today.

"Modern civilization is the daughter of coal, for this offers to mankind the solar energy in its most concentrated form; that is, in a form in which it has been accumulated in a long series of centuries," he told the gathering. "Modern man uses it with increasing eagerness and thoughtless prodigality for the conquest of the world and, like the mythical gold of the Rhine, coal is today the greatest source of energy and wealth. The earth still holds enormous quantities of it, but coal is not inexhaustible. The problem of the future begins to interest us."

The same patterns hold for hydrocarbons today. The more we dig for oil and gas, the shorter in supply it is and the more expensive it becomes to find. Our oil-based economy has made some countries very rich; the buying and selling of it dictates political alliances and rivalries around the world. The "thoughtless prodigality" that Ciamician saw then continues today, leading to an unprecedented and irreversible alteration of the earth's ecosystems that is leading to what some call the "Sixth Extinction"—an event on par with the five great cataclysms that destroyed vast swaths of life on earth, including the asteroid that killed the dinosaurs some 66 million years ago.

"Is fossil solar energy the only one that may be used in modern life and civilization?" he asked. "That is the question."

The answer, he believed, lay in the limitless power of the sun.

"Even making allowances for the absorption of heat on the part of the atmosphere and for other circumstances, we see that the solar energy that reaches a small tropical country—say of the size of Latium—is equal annually to the energy produced by the entire amount of coal mined in the world! The desert of Sahara with its six million square kilometers receives daily solar energy equivalent to six billion tons of coal!"

Ciamician was making the same argument that solar energy advocates make today, though the numbers are slightly different. Even today, the energy in the sunlight that hits the earth's surface in an hour exceeds the global human population's energy use in an entire year. Solar energy could meet all of humanity's growing energy needs—if only we knew how to make good use of it.

A passionate and vigorous scientist, Ciamician could deliver a 1.5-hour lecture with such force that "at the end of his lecture he was received by an assistant with a heavy wrap—much as our modern football players who retire exhausted from the game," marveled a researcher who visited him in 1913.

The scientist's roof was in many ways the heart of his lab—he lined tall, swan-necked glass tubes along its edges, like an array of beautiful, orderly stalagmites, and when the light hit them they must have given the building an otherworldly air. Perhaps this lab inspired his vision for the future.

"On the arid lands there will spring up industrial colonies without smoke and without smokestacks; forests of glass tubes will extend over the plains and glass buildings will rise everywhere; inside of these will take place the photochemical processes that hitherto have been the guarded secret of the plants," he wrote.

The scientist's vision was not unfounded. Nearly thirty years before his ambitious predictions of a world powered by the sun, American inventor Charles Fritts introduced the first solar cells in 1883, made from the semiconductor selenium. Fritts believed that his solar cells could compete with the coal-fired power plants that Thomas Edison had recently begun setting up to light customers' homes, but he was stymied by his device's low efficiency of 1 percent, due to selenium's imperfect characteristics. (In any case, history shows it did not pay to go toe-to-toe with Edison; the man was a brutal businessman who patented his employees' inventions, drove the nascent movie industry out of New Jersey all the way to what would become Hollywood, and mounted a successful smear campaign against electro-genius Nikola Tesla—after whom the famous electric-car-making company is named.)

So Ciamician was right about those tall glass buildings, though not for the reason he had hoped. The skyscrapers that mark cityscapes are monstrously energy inefficient, and the electricity that powers those cities—a clean-seeming power source—is still primarily produced by burning coal.

Learning the guarded secrets of plants is not enough to usher in a new energy age, when the vision comes at a higher price than the current flawed-but-cheap option. It's the kind of massive evolutionary shift that, for better or worse, may only come through a crisis. After all, it was the violent extinction of the dinosaurs that left vacancies in previously occupied ecological niches, allowing for a dramatic diversification among a previously unassuming group known as mammals that led, ultimately, to us. The question is, when the next environmental catastrophe hits, who will profit from the evolutionary roulette, and who will ultimately suffer?

For the artificial leaf, one such crisis occurred four decades ago from 1973 to 1974, when the Organization of Arab Petroleum Exporting Countries placed an embargo on oil exports to the United States, in response to the U.S. involvement in the 1973 Arab-Israeli war.

It's a shared trauma for most solar-fuels researchers I've spoken to: long lines at the fuel pump, gas rationing by license plate, an economy dragged down by the hiking of prices that brought oil to $12 per barrel, roughly four times the previous going rate. That makes sense, given that many researchers currently enjoying the prime of their careers are the ones who came of age during the early seventies.

But the embargo, and the oil crises that followed in subsequent years, were also an opportunity for clean-fuels researchers, leading to an unprecedented diversification in energy research. The crisis ushered in the creation of the U.S. Department of Energy in 1977, the establishing of the National Renewable Energy Laboratory, and a sudden influx of research funding for clean fuel, including hydrogen. The panic led to decisions made with incredible foresight, because it forced Jimmy Carter, the leader of the free world, to come to grips with the idea that we, as a society, had to cut our reliance on oil. Granted, the decisions were not motivated by the highest-minded ideals of protecting the planet from further damage, but by political supremacy and economic stability; still, they provided the soil in which better ideas could take root.

This sudden interest in renewable fuels did not last long; soon after the embargoes fell, oil prices stabilized and the world began to return to business as usual. Ronald Reagan, elected in 1980, took down the solar panels that Carter had had installed on the White House roof. Much of the funding for hydrogen research dried up, and many scientists moved on to more lucrative fields. But a few have kept on over the intervening decades, though their diverse efforts have increasingly fallen into two camps: those using the semiconductors used in highly successful technologies like solar panels, and those who want to use organic molecules in systems that are closer analogues of the natural system inside plants. In recent years, their collective work has begun to bear fruit—even as they continue to face major technological and practical obstacles.

One of the scientists inflamed by the oil crises of the 1970s was German electrochemist Heinz Gerischer. As a young man, he'd begun studying chemistry at the University of Leipzig in 1937 until his education was interrupted by World War II, pulling him into two years of military service. But Gerischer's mother was Jewish, and so in 1942 he was kicked out of the army, deemed "undeserving" to participate in the war effort. This allowed Gerischer to complete his degree in 1944, but those years were difficult ones for his family: his mother committed suicide in 1943 and his sister was killed in an air raid the next year.

Gerischer found support and inspiration after he began working with chemist Karl Friedrich Bonhöffer, whose family's anti-Nazi credentials ran deep: his brother Dietrich Bonhöffer, a Lutheran pastor and theologian, was executed along with his brother Klaus and two of his sisters' husbands for plotting to assassinate Adolf Hitler. It is in part because of Karl Bonhöffer that Gerischer was able to continue his studies after his expulsion from the military, because the chemist and others at the university reportedly helped conceal his Jewish ancestry. Bonhöffer also kindled Gerischer's interest in electrochemistry, a line of study that would ultimately lead him to a host of major contributions to the field, including pioneering research into semiconductors and their role in chemistry powered by light.

Like plants, silicon cells harvest the energy of sunlight by transferring the energy of photons into electrons, but that's basically where the similarity ends. Here's a very basic description of how they manage this trick: Take a crystal lattice made of silicon, with an atomic number of 14, that has 14 protons and 14 electrons (and 14 neutrons, but we don't care about them for the purposes of this explanation because they carry no charge). These electrons, depending on their energy levels, can occupy specific spots in some prescribed energy bands: two spots in the lowest band and eight spots in every band above that one. So silicon has 2 electrons in the lowest band, 8 electrons in the second band, and the remaining 4 electrons fill up half the empty spots in the third, highest band. That outermost, highest band is known as the valence band. For our purposes, valence electrons are the ones that matter because it's these electrons that can jump from the valence band into the conduction band, where—once channeled into an electric current—their energy can be put to use.

To make that jump, though, the electrons need a little extra energy—enough for them to scale the "band gap," the no-man's-land that lies between the valence and conduction bands. That energy comes in the form of photons. Light, as you might remember from high school physics, is both a particle and a wave: The particles are photons, little packets of energy that can bounce into an electron and transfer that energy into the charged particle, giving the electron the juice it needs to reach the conduction band. There, it's free to unbuckle its seat belt and move about the crystal cabin, free of its parent atom.

If you want an electrical current, though, you have to make these electrons flow in a circuit—you have to channel them. So scientists did this with silicon, using what's called a positive-negative, or p-n junction.

Silicon makes a beautiful crystal: Remember, it has four electrons in that uppermost valence band. These electrons are the ones that make covalent bonds—they share electrons with another atom so that together they can have the perfect eight electrons in their valence shell. Silicon hooks each of its four electrons to one electron from four other silicon atoms—and the result is a perfect interwoven crystal. Rendered in two dimensions, it looks like a tic-tac-toe board.

But perfection is useless, because in a pure silicon lattice, all the electrons are stuck in bonds and, because they can't move through the crystal, can't do any real work. So the scientists introduce some impurities into the crystal, a practice known as doping, in order to make the silicon conductive. They then sandwich together two slightly different pieces of silicon to make a solar cell. The first is p-type silicon that has been doped with boron; since boron only has three valence electrons, it leaves an empty spot, or "hole," in the crystal lattice. The second is an n-type silicon that's been doped with phosphorus, which has five valence electrons; that's one too many for the crystal to handle. Now, since the extra electrons in the phosphorus-doped silicon have lost the subatomic game of musical chairs—all eight slots in the valence band are already filled—they go wandering through the crystal. (Positive charges, or holes, "move," too: As one electron moves into a hole, it leaves another hole behind. Another electron then rushes in to fill that hole, leaving another hole behind. Because of this, you can think of holes as moving in the opposite direction as electrons. Holes are just the absence of an electron, which confuses me to no end.)

When you join the p-type and n-type silicon and connect them through an external wire, the p-type silicon pulls in extra electrons through the wire from the n-type silicon. Now the silicon on both sides feels good about this exchange, because all those empty holes and extra electrons have found mates. But the boron and phosphorus atoms are not happy at all—the phosphorus wants its electron back and the boron feels saddled with an extra electron it never asked for. This exchange creates a neutral "depletion zone" that electrons can only cross in one direction. This is what creates the electric field.

Now, when sunlight kicks an electron into the conduction band, that roving electron feels the attraction from the unhappy phosphorus atom and gets pulled over to the n-type. Of course, this means the boron-doped silicon in the p-type is now missing an electron. So it starts grabbing electrons from the external wire—and as the electrons travel along the wire from the n-type to the p-type, you can harvest the energy by running it through, say, a light bulb. Placed onto a roof and hooked up to your home's energy grid, these tiny particles can power your world.

Silicon isn't the only material used to make solar cells. Some use gallium arsenide or cadmium telluride. Others use amorphous silicon, a type that is less efficient because it lacks a crystalline structure, but it's not as expensive to make and can be shaped into thin, flexible panels. There are dye-sensitized photovoltaic cells, less efficient but also made of less expensive materials (and exciting because they can potentially be cheaply printed and laminated onto the sides of buildings, though the efficiencies might be too low for bulk power generation). Still others eschew inorganic materials, choosing to focus on understanding the complex electrochemical processes happening inside of a natural leaf that manage to trick photons to push electrons around to make usable chemical products. After all, the thinking goes, nature has been able to perform the same feat using organic compounds at room temperature, while your typical solar panels require highly purified (and thus, expensive) semiconductors to work.

"They're just different approaches to do the same job overall," said Devens Gust, an Arizona State University chemistry professor whose research has leaned toward building more organic, truly leaflike artificial systems. "What people are trying to do is figure out which is going to be the best. So far, the only system that's really practical is photosynthesis, and that uses the molecular system."

It's hard to argue with him on that point; inefficient though it may seem, nature has developed the most robust, long-lasting system to date, without needing harsh, expensive chemicals or extreme heating or cooling. Be that as it may, silicon and other semiconductors have several decades' and countless dollars' worth of leg-up on the competition and the technology that currently holds the best chance of making it to market. So, for the purposes of this chapter, I'll be focusing primarily on its development. (In any case, differences in technology seem to lie mainly on the light-gathering side. If the right water-splitting and carbon dioxide-reducing catalysts are found, they could aid both the organic and inorganic efforts to develop commercially viable artificial photosynthesis.)

In the 1960s, scientists were researching the potential of photovoltaic devices, which were expensive and difficult to make. In order to create a p-n junction, the researchers would have to take an n-type semiconductor and carefully dope a thin layer on top with the p-type atoms (or take a p-type semiconductor and dope the thin layer with n-type atoms). The problem is, these dopants would often sink deeper into the material than they should have, ruining the junction. It was a tricky and expensive process. Gerischer realized you could cut out much of that hard work by instead creating an interface between a solid semiconductor (either p-type or n-type) and a fluid electrolyte.

"Gerischer saw that semiconductor liquid junction cells would be simpler to make than p-n junction cells, because their junction is spontaneously formed when a semiconductor is immersed in a redox couple solution," Adam Heller, a chemical engineer then at Bell Laboratories in New Jersey, wrote in a 1981 paper on the development and evolution of the photoelectrochemical cell. (A redox couple solution is basically a solution full of complementary ions: some that can be reduced and some that can be oxidized. Hence the term "redox.")

Gerischer's impact on a generation of colleagues in his field was wide and deep; one fellow scientist described his number of coworkers as a function of time as "an exponential curve." The 1977–1978 academic year he spent at the California Institute of Technology—now home to the Joint Center for Artificial Photosynthesis—was no exception.

"That is indeed who taught me photoelectrochemistry," says John Turner, a well-known solar-fuels researcher at the National Renewable Energy Laboratory (NREL) in Golden, Colorado. Turner had studied batteries as an undergraduate and techniques for making electrochemical measurements as a Ph.D., but his postdoc at Caltech overlapped with the German scientist's brief stint on campus. "That's where I learned the field, was from Heinz Gerischer."

Turner, armed with this newfound love of photoelectrochemistry, joined NREL (then known as the Solar Energy Research Institute) in 1979. Using electricity to split water, or electrolysis, had been a well-known phenomenon; a child (under supervision) could do it with the help of two pencils, two wires, a cup of water, and a 9-volt battery.

But the idea of harnessing the sun to generate the electricity to make hydrogen fuel really took off in 1972, when Japanese scientists Akira Fujishima and Kenichi Honda demonstrated in a paper in *Nature* that titanium dioxide, a humble ingredient found in paint and sunscreen, could absorb light and actually split water. (This discovery, as it happens, earned Fujishima the inaugural Heinz Gerischer Award in 2003.) While titanium dioxide never took off for several reasons—largely because it could only absorb a tiny slice of sunlight in the ultraviolet range—the timing, a year before the oil embargo, was what one *PNAS* paper called "impeccable."

"That sort of opened up a huge field that I became a part of," Turner said.

Water-splitting's industrial use goes far beyond transportation fuel—hydrogen is a key ingredient used to refine food oils, as well as to manufacture ammonia, the stuff in fertilizer that allows us to feed an extra three billion people on the planet, Turner pointed out. Even that hydrogen is primarily obtained by a cheaper method known as steam reforming, in which the hydrogens are stripped off of methane gas and carbon dioxide is released into the air. This process is far from clean: for every one kilogram of hydrogen made with this method, twelve kilograms of carbon dioxide are produced. When you consider that more than fifty million metric tons of hydrogen are produced worldwide every year (including nine million in the United States), those carbon dioxide emissions really start to add up.

Scientists had hoped to clean up that process by making the sun produce that hydrogen, instead of culling it from fossil fuels. That process, photoelectrolysis, involves two steps: gathering photons and translating them into an electric charge, and then using that charge to actually split water into hydrogen and oxygen. Doing the two steps separately—gathering the charge in one place using semiconductors and then ferrying it through a wire to metal electrodes submerged in liquid—is one way to split water cleanly, but it ends up being a pretty costly system, Turner said.

"If you couple expensive electrolyzers with expensive photovoltaics, only running during the daytime, you can see how that gets very expensive," Turner says.

Researchers had also tried submerging the semiconductors directly in liquid, essentially making them do both jobs of light-gathering and water-splitting in situ. The problem, however, was that the submerged semiconductors quickly corroded, rendering the device inactive.

Turner managed to surmount these individual problems just enough to make a functional, effective device. In 1998, he and Oscar Khaselev announced that they'd created an integrated, wireless device that could take sunlight and convert it into hydrogen at a record-breaking efficiency of 12.4 percent. That was about twice as efficient as the best wired versions that separated the solar panels and the electrolyzer, and it's a record that he held for seventeen years, from 1998 to 2015.

While that is an impressive length of time for such a record to stand, particularly for a field on the technological cutting edge, it's also telling—and disappointing—that it took nearly two decades for someone else to break it. There are many issues, Turner himself will acknowledge, but one of the main challenges is finding semiconductors and catalysts that can last more than a day (Turner's degraded significantly over the twenty hours that they ran it) and that are inexpensive enough to manufacture on a large scale (Turner's used platinum and other rare elements).

Turner has since improved the system's lifetime, running experiments for one hundred hours or so, and they're working on taking that efficiency record back. But cost, in the end, is the main barrier to commercializing artificial photosynthesis. If the device that generates the hydrogen is too pricey to compete with oil and natural gas, there's little commercial incentive to make the switch, no matter how pressing the political or environmental necessities. Turner witnessed that himself in the 1980s, when the price of crude oil dropped and, in spite of the multiple oil-triggered economic traumas of the 1970s, funding for renewable energy research began to dry up.

"The United States's dealings with these risks were akin to the analogy of a smoker who, having perceived the risk of heart disease, takes up running—but continues to smoke," Turner wrote in *Science* in 1999.

To this day, the electrochemist is still a believer in hydrogen, in spite of the challenges that remain in building an efficient, commercially competitive device. After all, our fossil-fuel-based civilization has been around for a mere three centuries, and the planet is already suffering from the consequences. Turner wants an energy infrastructure that could last millennia.

Dan Nocera, a chemist at Harvard University, remembers well the aftermath of the oil crises of the 1970s, the spike in political interest in clean energy, and the ultimate funding crash in the decades that followed. He watched as many of his peers switched to other research fields.

"But I never went away from it," Nocera says, leaning back in his chair. "The rest of the world did, 'cause you all need to get your research funded . . . it was obvious to me that it would come back again someday."

Nocera is sitting in his corner office at the end of the dimly lit hallways in Harvard's Mallinckrodt building, whose formidable name matches its brick face and thick ionic columns. His office is large but spare, and I shuffle my chair closer to his desktop to get a better look at the diagrams he pulls up on the screen. The chemist has just returned from a conference in Australia on solar power and probably jet-lagged, but in between classes and a keynote speech, he makes some time to show me the lab.

If you've heard anything about the artificial leaf, then you've probably heard Nocera's name. The chemist made major headlines around the world in 2011 when he announced the creation of a device that could split water into oxygen and hydrogen gas using relatively cheap catalysts made out of earth-abundant elements such as cobalt and phosphate (on the side that makes oxygen) and nickel and molybdenum (on the side that makes hydrogen). When the sun hits the device, the positive "holes" generated by a triple-junction solar cell are sent over to the anode, where the cobalt-phosphate catalyst uses them to rip electrons off of water molecules, leaving behind oxygen (O_2) molecules and hydrogen ions (H^+). The electrons head in the opposite direction, over to the cathode, where the nickel-molybdenum-zinc catalyst sews those leftover hydrogen ions together to make hydrogen gas (H_2).

It's a beautiful little device: the size of a playing card, the triple-junction solar wafer can be dropped into a glass of water sitting in sunlight and bubbles of oxygen gas will immediately start forming on one side, bubbles of hydrogen on the other. In theory, the oxygen could escape into the air while the hydrogen would be collected and stored, in order to later be burned in an engine or used in a fuel cell. When consumed, the hydrogen fuel recombines with oxygen in the air, which means its only emission is water; there are no greenhouse gases of any kind.

The artificial leaf was neither the first attempt at a wireless solar-fuels generator nor was it the last, but it took both trade publications and the mainstream media by storm because it was so elegant: it required no wires and it could work in dirty water, which meant it could be hardy enough for rural conditions where clean water was not easily available. The catchy name didn't hurt, either.

I first spoke to Nocera in 2013, not fully realizing that he was calling me from a stairwell in a New Orleans restaurant on the night of his Kavli lecture to the American Chemical Society's national meeting, stealing a few minutes away from dinner with other presumably notable scientists to chat with a reporter on deadline. On the phone, two years after his groundbreaking announcements in 2008 and then 2011, he was still ebullient, energized, perhaps still riding the applause from the crowds of chemists at the meeting that day. I looked the lecture up later, to get a sense of his thinking that night.

For the MIT-turned-Harvard scientist, his research journey to building the artificial leaf goes back more than three decades. After the oil crises subsided and funding dried up, "energy went away, it was gone again. And I had no conferences to go to, isn't that sad?" he told the crowd in New Orleans. "All my friends became organometallic catalyst chemists, and I just kinda hung in there on energy."

Nocera's approach to hanging on was to do basic research that he thought might ultimately be useful to a number of disciplines, including—if he were able to return to it someday—mimicking photosynthesis. One of those processes, known as proton-coupled electron transfer, is key to understanding how plants manage to simultaneously perform the intricate dance of oxidizing water and reducing protons to hydrogen. Photovoltaic cells use sunlight to move electrical current, in the form of negatively charged electrons. But plants can convert that electrical current into chemical current, by making protons carry positive charge—even though protons are almost two thousand times heavier than electrons and harder to move around. Proton-coupled electron transfer offered insight into that process by describing "how electrons talk to protons," Nocera said, in order to keep the intricate dance of photosynthesis running smoothly.

"Everybody was looking for the magic material that would be the semiconductor—absorb the light—and split water," Nocera said, in part thanks to the 1972 Honda-Fujishima discovery. But in 2001, even after developing a rhodium-based catalyst that generated hydrogen from an acidic solution, he realized that this substance probably didn't exist. For one thing, it's hard to get a catalyst to absorb a wide-enough band of sunlight to be really useful. For another, water-splitting requires four electrons at a time, but semiconductors usually produce them one by one, which is another reason titanium dioxide had poor overall efficiency. So Nocera eventually turned back to silicon (a tried-and-true technology) to produce the electrons from sunlight, while searching for the right catalysts that could grab four of those single electrons and do the water-splitting work.

"You're never going to make a magic material, just like you're never going to make a magic molecule," he said. "Because I'd already been around that block."

In some ways, divvying up those two tasks was taking another page out of nature's book, Nocera said—plant cells also perform light-harvesting and charge separation separately from catalysis. Still, many of the scientists working in semiconductor-based solar fuels eschew the term "artificial leaf," because to them, the process they're trying to develop is nothing like a leaf. They have a point: plants use organic molecules to shuttle electrons through various stations in a watery, mostly neutral-pH environment, making those electrons do work along the way, before ultimately locking hydrogen ions (also known simply as "protons") into carbon-based molecules called sugars. They're inefficient, converting only about 1 percent of the light that hits their leaves' surfaces into energy, and they're unstable processes. That's nothing like solar-fuel generators, which require highly purified inorganic material like silicon to skip the sugar-making and split water directly, and at a much higher efficiency than is possible in nature.

The chemist doesn't deny that the artificial leaf doesn't operate using the same mechanisms as biology; his goal is to translate the function of the leaf into a working device. Of course, to do that well, you do have to study how nature does it.

"The way you get a deep, deep understanding is with molecules, because you can define everything," he said. Once you do that, you can apply those lessons using inorganic materials, under different conditions, for a different end product.

Nocera still sees many other parallels between nature and solar fuels that he says have informed his work. In his office, he pulls up an illustration of the innards of Photosystem II, the first complex in the plant to absorb light and use it to oxidize water. It's a twisty, Technicolor rendition of entangled proteins; a blue squiggle represents the D1 protein, which holds a manganese-containing compound known as the oxygen-evolving complex. This complex does the hard work of taking two water molecules and stripping off the four hydrogen ions so that the two oxygens can come together and form O2, or neutral oxygen gas. This process, known as oxidizing water, is not easy—in fact, it damages the protein holding the complex in place. In that way, it's not all that different from the trouble with using silicon wafers to oxidize water—the silicon gets oxidized; that is, it rusts. But in plants, the system is used to that kind of wear and tear; it actually replaces the protein with its worn-out ligands that hold the manganese molecule about every thirty minutes.

"That's happening in every plant," Nocera says.

That self-healing process is a property Nocera sees in his own water-oxidizing catalyst, a cobalt ion with a phosphate buffer. The researchers think that when they run an electric current through the device, the anode sucks away more electrons from the already-ionized Co^{2+} floating in the liquid, turning them into an even more ionized form, Co^{3+}. This precipitates out of the liquid and joins with phosphate ions on the semiconductor's surface. At this point, the current then pulls another electron off of the cobalt, making it Co^{4+}. This super-ionized form can then rip electrons off of water molecules, stripping the hydrogens and leaving the oxygens to join together and make O_2 gas. In this process, the cobalt ion gains electrons, jumping up to Co^{2+} and breaking off from the surface to float in the liquid, and the process starts all over again.

In 2008 Nocera and his coauthor, Matthew Kanan (now a professor at Stanford), accidentally stumbled upon this compound's properties; they had tried it out for kicks, layering it on top of their semiconductor wafer, and assumed the attempt had gone south when a greenish layer began growing on the surface. But after oxygen bubbles began forming on the chip and the researchers examined the catalyst, they realized they had a special compound on their hands—one that could break down after reuse and then, with a little energy input, reassemble over and over again. This, to Nocera, is a perfect analogue of the plant's self-healing manganese complex.

This discovery was crucial, because oxidizing water—stripping out electrons, leaving a bunch of protons floating around—is the more difficult, more complex side of the reaction. Nocera put a different but also-cheap catalyst made of nickel, molybdenum, and zinc on the hydrogen-producing side and in 2011 his artificial leaf was set.

The chemist doesn't show me the device itself. This is late 2014, well past the hype and the magazine profiles and the well-produced video segments that featured him holding actual leaves, sun glinting in the background. Even as Nocera's device brought him international public attention and invitations to speak at chemistry research meetings and institutions, he was attracting criticism from some colleagues for making promises and prophesies for the future that his current technology could not meet.

"Another term for self-healing is 'unstable,'" Turner pointed out. "You don't need self-healing if something's stable. And if something's unstable, there's always the possibility that something won't self-heal."

Keep in mind, the artificial leaf was not necessarily the most efficient solar-fuels device out there: the device operated at 4.7 percent in a wired configuration, where the solar panels and the electrolyzer were linked by a wire; and only about 2.5 percent in its integrated, more leaflike form. Nocera showed that it could be done more cheaply. However, the efficiency of the water-splitting device is tied to the semiconductor—and until those get better and cheaper to manufacture, costs will stay up. (The artificial leaf contains a triple-junction solar cell, which essentially sandwiched three photovoltaic semiconductor systems together to capture different wavelengths of sunlight. While that improves its sunlight-to-electricity efficiency, it means using roughly triple the material, making it slightly more expensive from a materials perspective, too.)

The company Nocera founded in 2008 to commercialize the artificial leaf technology, Sun Catalytix, earned venture capital funding from Indian conglomerate Tata as well as the Advanced Research Projects Agency-Energy, a Defense Department arm that seeks to fund cutting-edge technology. But the edge that cuts just as often draws blood, and Sun Catalytix's assets were sold to Lockheed Martin in mid-2014, a few months before my visit, without getting close to a commercial product. The company then dramatically changed direction, focusing instead on developing flow-battery storage, apparently a more attainable technology in the short term.

"VCs work because of this stupid thing, right?" he says, waving toward his monitor. "Computers. And apps. It's low capital-intensive—it doesn't cost a lot of money to have people in their basement writing software. But energy is capital intensive. And you have big energy companies—you're competing against them."

Nocera's point is this: Even if the artificial leaf experiences a massive leap in efficiency and the expense manages to fall low enough for it to compete with hydrocarbon fuels, there's no good way to store hydrogen fuel on a large scale. Hydrogen is the lightest element on the periodic table; the gas it makes has to be extremely compressed, which requires a whole new kind of storage technology and energy infrastructure. Our admittedly aging energy infrastructure has been built up over the intervening century and a half; it's economically unfeasible to build out a new grid in one go. (If you're asking, why not just have everyone make their own hydrogen? That's a good question. It may or may not be economically viable for everyone to have a hydrogen generator and fuel-cell electrolyzer in their homes, but either way, the issue is moot because of legacy: ever since Thomas Edison's first coal-fired power plant made centralized power cheap and convenient, we've been wedded to the grid.)

"I really do think the model I developed is the right one," Nocera said. "Which means, if you're in this business to do the research and really, like, change things, with a vision, you have to put greed second. Because I can't do it with thirty people and 110 million dollars."

Nocera still thinks his hydrogen-producing artificial leaf technology would be useful for developing countries, where there's little to no existing grid and where having an independent source of energy could dramatically improve the lives of those in rural areas, extending their day into the dark hours, allowing children to study and adults to work their way out of poverty. In the United States, that doesn't seem to be happening anytime soon. So he's teamed up with researchers at Harvard's Wyss Institute for Biologically Inspired Engineering, designing systems where they feed hydrogen (presumably made with an artificial leaf of some kind) to genetically modified bacteria that use it to make isopropanol, an alcohol that can be used as a conventional liquid fuel, as well as some precursors to bioplastics. Coupled to existing photovoltaic technology, the device can reach efficiencies of about 10 percent. Nocera calls it the bionic leaf.

"I just wish the world would start using hydrogen as a fuel but I can't convince the world," he tells me. "So it has to get messier, to adapt to the current energy infrastructure."

The Caltech campus sits northeast of downtown Los Angeles, set against the hazy blue-green backdrop of the San Gabriel Mountains. It's early spring 2015, and even though some trees seem to be suffering the consequences of Southern California's drought, there are plenty of green canopies shading the walkways cutting through the grassy campus. That's a relief, given that the sun always seems to beat down a little harder here in Pasadena than it does in downtown.

It was one such sunny day that Nathan Lewis, then an undergraduate student in Harry Gray's lab, had been walking through campus, carrying a sample he'd prepared containing a rhodium complex to another building containing a nuclear-magnetic resonance machine, where he'd analyze the physical and chemical properties of the atoms within. Half the tube was safely ensconced inside his pocket and the other half stuck out, exposed to sunlight. When he pulled it out, he saw that the sunlit half had turned from bluc to yellow, and had generated a gas—which turned out to be hydrogen.

"Oh, it was great!" Lewis says of the accidental discovery, as we walk through campus. "It was the first kind of foray into solar fuels, really, here at Caltech, and it was during the oil crisis in the seventies—you know. It was a fun time."

Rhodium itself never made much of a splash after that—the metal is not fully catalytic, as it happens—but the seed of an idea had been planted. In the mid-2000s, after a career spent on other pursuits (including the electronic nose), Lewis gathered his group on a retreat, where they discussed the next direction to take their research. It was around the same time that he'd gone on a family vacation to Hawaii—an annual pilgrimage—and his thirteen-year-old son had asked why the coral at the same beach they'd visited the year before had been bleached. Lewis explained as best he could: the carbon dioxide we were pumping into the air was sinking into the seas and "fizzing the oceans," making them acidic and unlivable.

"Can't we stop it?" his son asked.

"Not unless someone finds a way," Lewis said.

"Well," his son asked, "why don't you find a way?"

During the retreat, the researchers drew a rough sketch of a device they might build. Instead of flat panels, they drew two arrays of semiconductor microwires—tall and needle thin, reminiscent of a miniature forest. The idea was this: As soon as light generates a free electron in gallium arsenide, silicon, or any other semiconductor, the electron has to travel all the way back through the material to reach the wire to generate a current—a dangerous journey, because it gives the electron time to recombine with a positive "hole" and never make it into the wire to generate that much-needed electric current. But if you make microwires, you allow for a long surface area on which the semiconductor can absorb light but a short hop—across the needle's narrow diameter—that the electron needs to go to hit the catalyst. It's a geometric solution to an electrochemistry problem.

"The reason I had the idea was bioinspired—we wanted things like aspen trees that could absorb light over long distance but move the excited states a short distance sideways and not all the way back," Lewis explains. "That's why solar cells have to be so expensive, because you need a certain thickness to absorb the light, but then the excited electrons have to go all the way back the way they came, to get to the electrical wires. If it's less pure, they'll just make heat along the way and they're lost."

Just as forests grow vertically to house a whole lot of light absorbers, these microwires, with a significantly higher surface area, could serve as far better harvesters of light than a flat surface. And while I'm not sure Lewis knew this at the time of our conversation, it turns out microwires might be more like aspens than he thought: beneath their white bark, along their entire length, the trees have a thin green photosynthetic layer that they use to absorb light and generate fuel in the unforgiving winter months when their leaves have fallen.

Out of that, the NSF-funded Center for Chemical Innovation in Solar Fuels was born; it served as the basis for the Department of Energy–funded Joint Center for Artificial Photosynthesis (JCAP) established in 2010. JCAP, which received around $122 million from the DOE over a five-year period, has its home base on the Caltech campus, a satellite campus up at Lawrence Berkeley National Laboratory, and a collection of individual researchers located at several other California universities. The goal was as ambitious as it was straightforward: create a working device that could cheaply and reliably produce fuels using sunlight. To Lewis's mind, too many of the advances made in our understanding of artificial photosynthesis until then had been piecemeal. What breakthroughs had happened, in a specific catalyst here or a semiconductor there, weren't useful unless they could be integrated into a larger system. JCAP set out to build that system, "soup to nuts," the scientist said.

The chemist leads me into the Jorgensen building, which houses JCAP—a white building with an enormous glass façade that lets in lots of sun. The minimalist interior inside has a futuristic feel: the walls, floors, ceilings, and staircases are all white, reflecting back all the light from those large glass windows. See-through Plexiglas appears to hold up the metal guardrails, and the semi-open floorplan allows people walking along the second floor to see any and all activity in the lobby below. This place was designed by Lewis and a colleague, and it looks very different from the dark, narrow hallways and closed-off lab rooms of the chemistry labs I've seen.

"Yep. I wanted this lab to be all one floor but we didn't have the footprint to do it," he says as we climb the stairs, "because I wanted all the interaction to happen with all the different projects. We kind of get it because that big conference room is on the first floor . . . and we have food."

We put on clear safety goggles, conveniently placed in bins attached to the walls in front of laboratory doorways. Inside, each lab starts to look more recognizable, in that they hold a bevy of large scientific instruments, some of which look oddly familiar. There's a reason for that: several of these high-tech tools are repurposed inkjet printers, whose inks have been replaced with experimental compounds.

"They squirt out different inks from all the different elements we care about; every spot has a different set of colors that we coopted, a different set of inks, and then we anneal them all to make them oxides," Lewis says, over the machinery buzzing in the room.

He picks up a large slide with rows upon rows of printed dots. They're really kind of pretty—each dot a slightly different hue, like a tiny pantone palette, each shade indicative of a different chemical composition.

"They're beautiful. That has eighteen hundred spots on it, and every one has almost two thousand different compositions," he tells me. "We can print a million compounds before lunch."

This is essentially a high-throughput facility, searching for the right materials with the ideal characteristics for a solar-fuels device. And this printer is just one of at least three different methods used to produce chemicals whose properties will be subjected to a host of different tests: probed by tiny electric wires to measure the current, bombarded with X-rays to reveal their structure, subjected to a scanning robot that measures how much light a compound reflects (bad) and how much light passes through (good, so that light can reach the semiconductor underneath).

Some of the chemicals are designed—the researchers predict what a given compound's chemical properties may be, given what's known about the elements within them—but some are random surprises. Either way, if there are Holy Grail catalysts out there, Lewis plans on finding them.

"None of the things we've discovered yet have done as well as the things we've thought about and designed first," he said. "But that's got to surely change because we can't be that smart."

Lewis lists off the five basic things they need to make a complete system: Two catalysts, one for oxidizing water and the other for taking the resulting hydrogen ions (the leftover protons from the first reaction) and turning them into hydrogen gas; two semiconductors, one to capture the blue wavelengths of light and another to catch the red; and a membrane that can separate the two sections while also selectively passing the right ions through. (For example, a membrane that can pass only the protons generated at the photoanode through to the photocathode side, where they'll pair up and grab some electrons to make hydrogen gas.)

"It's not the case that a breakthrough was going to be made in a magic catalyst or a magic light absorber, and this whole thing is going to work, because you have to have a lot of these pieces. They all have to work together at the same time. It's like building an airplane—just because you have an engine doesn't mean you have a plane that flies," he said. "You need the wings, you need designs, the plane has to be aerodynamic, it's got to get off the ground—so it's much more. And when we had this vison, we had none of these pieces available. We didn't have the ways to make microwires, we didn't have the ways to orient them toward the sun, we didn't have the membranes, we didn't have the catalysts—we didn't have any of the pieces by themselves, no less working all together."

On top of that, the reaction can't take place in neutral-pH solution: it has to be either in an acid or a base, for the chemistry to work. That's actually a big issue Lewis has with Nocera's wireless "artificial leaf"—the idea of performing the reaction in neutral water doesn't work for long, because the very act of splitting water changes the pH of the water near the semiconductor surface. Liberating protons on the oxidizing side makes the local liquid more acidic; consuming protons by turning them into hydrogen gas on the reduction side makes the water more basic.

"The catalyst was designed to work in neutral pH and no system can stay there for very long," Lewis said, referring to it as freshman chemistry. "The only way to make a system that you can sustain, and this is known from a hundred years of electrolysis work, is to work in either a local pH that's acidic or a local pH that's very basic."

(Nocera, on the other hand, argues that highly acidic or basic solutions are corrosive, and that the best way to make a sustainable device is to operate in neutral conditions.)

Having a membrane between the photocathode and photoanode compartments is also key to the design, Lewis says; the gases must not combine, or they'll explode. That's another major issue he has with the "artificial leaf"—there was no clear method to safely separate the two gasses as they bubbled off the catalyst-coated surfaces.

"We never called it an artificial leaf," he said of JCAP's efforts. "That's kind of a jargony name that is fine, but it's not a leaf, it's not a living system, it doesn't look green, it doesn't uptake carbon dioxide to make sugars; it's a solar-fuels generator. It doesn't look like a leaf any more than a bird looks like an airplane."

Between JCAP and CCI Solar, Lewis says they've recently discovered an entirely new family of catalysts that are the best at reducing water to hydrogen, and they're made from cheap, readily available elements.

"We chose to publish them in peer-reviewed journals and not have chest beating press releases about them," he adds—a comment that strikes me as a reference to the media attention surrounding publications, like Nocera's, in *Science* and other journals. "We could have. We didn't."

JCAP also has a benchmarking facility where they test the long-term viability of catalysts made by other scientists. Most of them, Lewis says, do extremely poorly. "Even the cobalt?" I ask, in implicit reference to one of Nocera's catalysts.

"Dies in a day. It's gone," Lewis says. "So you see those demonstrations, you know, but I mean—who wants one day, right?"

Though Lewis and Nocera hardly if ever mention each other by name in the conversations I've had with them, and though they've even coauthored articles on the future of solar fuel, the way they refer to the other's research gives me the feeling that the two didn't see eye-to-eye much, if at all. It's a strange sibling rivalry of sorts; Lewis studied under Harry Gray as an undergrad; Nocera did so as a graduate student (though not at the same time). Both credit Gray with helping to inspire their interest in artificial photosynthesis.

"Even though they are quote, 'academic brothers' . . . they kind of disagree on almost everything," another solar-fuels researcher told me.

JCAP has been looking to build a solar-fuels generator in both acid and base, because they're not yet sure which system will end up being best. Each one has its own advantages. For example, there are cheap catalysts available for a basic-pH system. But at the photocathode, silicon would be stable in acid. Lewis says they think they have all the five pieces necessary to make a system in base; in acid, they're missing one.

A few months after my visit, Caltech announced that JCAP scientists had indeed built what the press release called an "artificial leaf"—a device that could split water in a base (in this case, a solution of potassium hydroxide) at a 10.5 percent efficiency rate. That's in spitting distance of John Turner's record, and it's a feat they managed to keep up for forty hours straight; the device's performance began to degrade around eighty hours. (They've since gotten it to run for hundreds of hours, Lewis told me a year later.)

The researchers, it seems, may be heading in the right direction. But right around the time of my visit to JCAP, the energy hub's five-year agreement had run out, and its progress and purpose had come up for review by the Department of Energy. JCAP's great appeal was its mission to develop a solar-fuel generator from start to finish, bringing a host of different researchers working on different angles of the problem under one roof. Now, its funding was to be slashed roughly in half, and its members would no longer be afforded the job security of a five-year time line. At that point, there were about 130 people working under solar fuels at Caltech under JCAP, a number that would now have to be cut, Lewis said. But roughly 100 more Caltech researchers work on solar fuels separate from the center, he added. Solar-fuels research would continue—though perhaps not with the same concerted effort.

"Figuring it all out, that was a unique part of what JCAP did and we hope to preserve that," Lewis said. "But with our budget cut, we don't know how much of that we're going to preserve."

On top of that, the center was given a new, much more difficult directive: switch from designing a hydrogen generator to a device that fixes carbon dioxide to make hydrocarbon fuels. While there are obvious advantages to producing such a liquid fuel—as Nocera pointed out, they're easily stored in conventional cars and existing energy infrastructure—it is a much harder task, because you have to string together many more electrons and protons to make a working fuel.

"There's debate about whether or not that's the optimal way to do this but that's what's going to happen," Lewis said. The plan, for the moment, is to continue to pursue both; after all, you probably need to know how to make hydrogen if you're also going to make a molecule out of carbon, hydrogen, and oxygen.

"We have unfinished business, right?" he says before we leave the lab. "There's no question, we have unfinished business."

There is a large painting of a tree in Peidong Yang's office, with wide, inviting branches under a broad canopy. Among the cases of textbooks and shelves of awards, it's a whimsical, eye-catching centerpiece.

"That's my daughter's painting," the UC Berkeley chemist says. "Actually, half-half. This half is my wife's, that half is my daughter's."

It's a beautiful piece, clearly made with love, and I'm a sucker for its jewel-tone hues. But the art strikes me as remarkably scientifically apt, given the direction of Yang's research: he has built a "synthetic leaf" system that's essentially a hybrid of two different processes—inorganic semiconductors and living, breathing bacteria.

When Yang won the MacArthur genius grant in 2015 for his work with nanowires, photonics, and solar fuels, I immediately kicked myself. I'd been told by other researchers that he was an important person to talk to in the field, but I'd put it off. Now, with all the other media attention, I was sure he wouldn't have the time to speak for a book what with newspaper and TV interviews. To my surprise (and relief), he invited me up to the campus to check out his laboratory and talk about his research.

"If we want to mimic what's going on in the leaf, to learn from natural photosynthesis, then at least we need to understand what's going on in the leaf," he tells me. "In the leaf, there are two catalytic cycles—one is for the water oxidation, another one is for CO2 fixation. That's the basic chemistry going on in natural photosynthesis. So if we want to design some inorganic material to do the same job, essentially the chemistry or the physics should be similar."

Unlike Nocera and JCAP, who have recently pivoted to liquid (or hydrocarbon) fuels, he's been working on the basic science of generating liquid fuels for a few years. Yang didn't really start out planning to generate solar fuel; he joined Berkeley in 1999 to work on better ways to make nanowires—arrays of thin, needlelike crystals that are about a thousand times thinner than a single human hair. The laws of physics start to change on a pretty fundamental level at those scales; materials, responding to quantum effects, start to behave in very unusual ways. Nanowires are so tiny that they can be smaller than the wavelength of light—a property that allows the nanowires to potentially channel individual photons with much higher efficiency than the same semiconductor would in its flat, bulk-material form. Yang was interested in nanowires from a basic science level, exploring their properties and their potential application in a wide variety of fields. He's examined their impact on solar cells' light harvesting abilities, turned them into tiny photon-shooting lasers and built arrays that could suck up waste heat and convert it back into electricity. (So much of the energy we produce—whether in power plants or by our warm-blooded bodies—is actually lost in the process as heat. Imagine if you could recapture that and put that electricity back into the system—you could make power plants far more efficient and use the heat from your car engine to power the radio and other electronics. On a smaller scale, you could use your own body warmth to charge your phone and other personal gadgets.)

Yang was thrust fully into the solar-fuels scene when former U.S. Secretary of Energy Steven Chu, then director of Lawrence Berkeley National Laboratory, launched the Helios Project with the aim of creating carbon-neutral fuels. The chemist soon realized that working on the semiconductor side wasn't enough to address this pressing energy problem: He would have to split his laboratory in two parts, one to work on semiconductor nanowires and the other to work on the catalysts that would coat those surfaces. Out of the thirty-five or so researchers in his group, about half are working on nanowires and half on the catalysts, he tells me, keys jangling heavily as he opens the door to his lab.

We walk past a lab bench lined with four or five heating plates that look like fancy postage scales; on top of each one sits a high-walled petri dish full of some clear solvent. He points to two doors, side by side, where they've conveniently split up study of the synthetic leaf's two main functions: turn light into electricity, and use electricity (with a good catalyst) to split carbon dioxide. "This is the light source," he says pointing to the lamp in one room, "and then the next door is the dark side. This side is no light; it's just doing the electrochemistry." Different catalysts can turn carbon dioxide into different products, he explains.

Another room holds the laser lab, its window carefully obscured with black trash-bag plastic and green tape—a safety precaution in case any beams somehow go rogue. They have cryostats in here that can bring the temperature of a material down to 10 Kelvin above absolute zero, he adds.

"Why would you want to do that?" I ask. Shouldn't these materials operate at room temperature?

"By studying temperature-dependent properties, we better understand the physical properties," he says. It's only in understanding how these nanowires act under extreme conditions that he can learn what they're really like. (It's sort of like seeing a road trip with a longtime friend as a test of character. Will she freak out when the car breaks down? How will I react after she's played that same awful song eight times in a row?)

Then Yang shows me the part of his lab that grabbed popular attention: the biolab, where he and his colleagues teamed up nanowires and bacteria to turn carbon dioxide into acetate, butanol, and even polymers.

"We have the incubator here, the culture, the media, those are where the bacteria are living," Yang explains, pointing at the device. "That's the apparatus where we have the electrodes; it's a liquid medium. And again, this is a solar simulator," he adds, tapping on the glass.

Light goes in; CO2 is bubbled into the water from a nearby tank. The bacteria, known as *Sporomusa ovata,* attaches to the semiconductor nanowires and grabs the electrons energized by sunlight, using them to reduce carbon dioxide to acetate, a versatile little molecule that can be used as a building block to a host of other carbon compounds. The acetate is then fed to genetically engineered *E. coli* bacteria, which take the acetate and turn it into a number of products, including amorphadiene (precursor to an antimalaria drug), butanol (used as a liquid fuel), or polyhydroxybutyrate (known as PHB, a biodegradable polymer that can be used to make plastic products).

Nanowires are key to both the light-absorbing and the catalyzing halves of the process, Yang says. Coupled with their strange quantum properties, they're also less reflective than smooth surfaces and don't let as many photons bounce off of the panel surface. The extra surface area allows for potentially more light trapping—and allows for more surface area for catalysis, by giving the bacteria more real estate to harvest electrons. That extra surface area is essential, because you don't have the option of spreading more panels over more real estate, as you would in a system where the electric charge generated by a typical photovoltaic solar panel is sent over to a separate electrolyzer. In a photoelectrochemical cell, the semiconductors are embedded inside a liquid-filled cell and have no room to grow. Besides, ideally you wouldn't want to spread more panels because that would take up more valuable real estate. Nanowires should help to make the most efficient device possible in the smallest space possible.

This work was a proof of principle; in terms of practical applications, the technology has a long, long way to go. The nanowires' solar energy conversion efficiency was 0.38 percent; the *E. coli* part of the reaction was more promising, with a conversion rate of 26 percent for butanol and 52 percent for PHB. And ultimately, Yang doesn't want to have to rely on bacteria to do the work of reducing carbon dioxide and turning it into something useful for humans. If researchers can design catalysts for oxidizing water and reducing hydrogen, then they should be able to do so for carbon dioxide. Unfortunately, that's easier said than done, because that molecule does not like being broken up.

"CO_2 is such a stable molecule," Yang tells me. "It's just—very difficult to crack."

And once you do break up a CO_2 molecule, its components are looking to date around. Those newly single carbon atoms hook up with any number and combination of oxygens and hydrogens to create a variety of different chemicals—not necessarily the ones that Yang is trying to make. Part of the issue is that once the carbon dioxide molecule is reduced, it probably has to go through several intermediate stages to reach a final product—a juggling act that's easy for a living cell, full of finely tuned enzymes for the job, but not so easy for a single catalyst. In the end, perhaps multiple catalysts will need to be developed to get carbon dioxide all the way from a greenhouse gas to a usable product.

"In terms of the weakest link in this whole system right now, it's the CO_2 catalyst," Yang says.

Since then, however, Yang and his colleagues have made progress; a 2015 paper in *Proceedings of the National Academy of Sciences* revealed that they'd raised the solar-to-fuel conversion rate (in this case, methane) to 10 percent—roughly twenty-five times more efficient than their previous work. A 2016 paper in *Science* showed that they'd "trained" a non-photosynthetic bacterium both to photosynthesize and to actually produce light-harvesting semiconductor nanoparticles by itself, a strategy that could help reduce the system's cost.

Yang understands the imperative here—and the urgency of finding a solution. Hydrogen is technically a cleaner fuel than a hydrocarbon—when consumed, it recombines with oxygen and the only waste product is water. But creating a solar hydrocarbon fuel comes with one big advantage: As you make it, you're actually pulling carbon dioxide, that greenhouse gas that's in large part responsible for global warming, out of the air. Even if you end up burning it, you've still created a "carbon-neutral" fuel—the fuel's life cycle doesn't result in any extra CO2 to the atmosphere. And if you make biodegradable plastic products instead of fuel, you're further ensuring that the carbon ends up in the ground, instead of the air or the increasingly acidic oceans. His synthetic leaf gives society the opportunity to solve two problems at once: generate fuel that doesn't require us to build out potentially trillions of dollars' worth of hydrogen infrastructure, and reduce our environmental impact at the same time.

"You don't need to dig from the ground anymore," he says, in reference to fossil fuels. "So we really need to work harder to make sure this thing will eventually work."

But the scientist isn't looking to rush a product to development. Change won't be coming anytime soon, in any case. "We will continue to dig for the next hundred years," he tells me. And sitting in his office with one leg crossed over the other, he looks remarkably calm about it all.

"I feel everybody is too anxious," he replies, when I comment on his surprising serenity.

I ask him about the recent changes at JCAP. As it turns out, Yang was the founding director of Berkeley's JCAP program, but left the helm after two years to focus on conducting scientific research. The recent changes at the center, which now have it focused on a more challenging problem, look promising to him.

"Philosophy wise, it's much calmed down," Yang says. "It's basically aligned with my philosophy. They are focusing on CO_2 catalysis. Focused on science."

Yang has two start-ups of his own. The research for the most successful one, Alphabet Energy, took ten years; developing that into a product on the market (a device that recovers waste heat and turns it back into energy) took another six. And that, by most standards, is pretty darn fast, he adds.

"I know how difficult it is to push from basic science to technology," he says. JCAP's first incarnation, he tells me, was "way, way too anxious to produce a so-called 'prototype.' My philosophy is, if you don't have the science, there is no technology."

This is why Yang continues to focus on carbon dioxide reduction, even though it's the most difficult reaction of the bunch to understand. For him, hydrogen reduction and water oxidation are essentially solved problems—the trick now is in refining and improving the processes and then integrating them into a device. In terms of basic science, understanding carbon reduction is still uncharted territory—and impossible to peg to a timetable.

"Scientific discovery cannot be planned," he says. "If you can plan a scientific discovery, then that's not a discovery."

8

CITIES AS ECOSYSTEMS

Building a More Sustainable Society

The Southern California Edison Energy Education Center sits in a low-slung office building in the city of Irwindale, framed by the San Gabriel Mountains to the north and flanked by the Santa Fe Recreational Dam to the west. East of the building lies the border with Azusa and the Cemex gravel-processing facility, a carved-up, barren landscape of dirt and rock.

Aside from the crown of purple-green peaks, there's little beautiful about Irwindale, I think as I drive past empty railways and parking lots. On this stretch of the eponymous boulevard, there are no homes in sight, not even a gas station or a convenience store. With a 2010 census population of 1,422, the city feels more industrial than residential, perhaps best known in recent years for long-running complaints that the Sriracha hot sauce factory owned by Huy Fong Foods was allegedly filling the air with spicy smells, leading residents to report burning eyes and asthma attacks.

Irwindale, in short, seems like a strange place to come to learn about the future of cities and the role of nature in their design. And yet here I am, in a conference room packed with hundreds of attendees, so many that for lack of seats I have to lean against the back wall. This is the inaugural Biomimicry LA conference, sponsored by consulting firm Verdical Group and featuring representatives from Los Angeles city government as well as consultants like Chris Garvin of Terrapin Bright Green, who's flown in from New York. Also attending: the U.S. Green Building Council, who announce that the next Greenbuild Expo will be coming to the City of Angels in 2016 for the first time in its fifteen-year history.

In the future, cities will be the greenest places to live. And in many ways, they already are. Manhattan's asphalt streets and piercing skyline may not, at first glance, make an impression as a particularly "green" human habitat, compared with the leafy suburbs of Tarrytown or the rolling farmland outside Rochester. But between the three, the city of concrete and glass may well be the most ecologically friendly human habitat.

That's because cities can end up being far more efficient utilizers of resources, including infrastructure, energy, and human capital. Urbanites often live in houses, condos, or apartments squished close together, which saves energy. They may be more likely to walk or bike or ride the metro to work, and even those who drive do so far less than those in rural environments. In the summer, city dwellers may turn the air-conditioning a little higher to fend off the "heat island" effect—but in the winter, all that extra heat keeps the gas bills low.

This concentration of resources also makes cities an increasingly appealing option. More than half of the world's population lives in urban areas, and that share is set to rise to 66 percent by 2050. Much of the population growth and population shift will occur in developing nations. Some of them, like China, have built and even overbuilt, creating cities for people with no one to fill them. But they certainly look more ready for the coming population shifts than India does. In 1950, 13 percent of China's population lived in cities, significantly less than the 17 percent of urban dwellers in India. But from 1950 to 2005, China's urbanization rate was 41 percent, far outpacing India's 29 percent, according to the McKinsey Global Institute.

Of course, that's not to say cities are perfect. The picture of whether they're environmentally friendly becomes more complicated when you look beyond measurements like per capita carbon footprint. Certain cities are better than others. It depends, among other things, on how much and how far residents have to drive to get to work, the state of public transit, whether healthy food sources are nearby, and what types of homes are available (apartment units are much more energy-efficient than ranch homes because they don't lose heat through all four walls, for example).

On top of that, cities have a bad habit of living beyond the means of their local environment. The food that cities eat is often grown elsewhere, taking energy and water from another place to produce, and that cost often isn't taken into account when researchers total up the environmental bill. So, too, is the potable water, pulled not from local sources but from rivers hundreds of miles away. Cities, in that way, have been pulling a colossal dine and dash.

That's not to mention that many cities are dealing with the realities of existing, aging, inefficient infrastructure that they simply don't have the money to replace all at once. That's why, in some cases, much of the interest in applying bioinspiration to whole systems is coming not from cities in the developed world, but from those in developing nations.

In those places, there are some organizations that do have opportunities to remake environments on such grand scales; St. Louis–based architecture and planning firm HOK is one of them. The firm helped to develop the first LEED (short for Leadership in Energy and Environmental Design) rating system in the

1990s, created by the U.S. Green Building Council. Making the so-called "built environment" sustainable has long been a core mission for the firm, and incorporating designs that are inspired by and work with nature is just one of the tools they use, says Chris Fannin, director of planning for HOK's Asia Pacific arm.

"We often get reminded by our clients here, 'Well, you're not talking about sustainability enough.' And we say, 'Well, you just get that. [You] don't have a choice,'" Fannin told me. That's simply part of the deal, particularly in their large-scale planning projects, where they have the opportunity to remake whole cities.

"The way we look at it is, the impact that we as a firm have on the built environment is significant," he added. "When you look at the potential impact of your work, the responsibility that comes with that is deeper than the project and the client. This has to be taken very seriously."

Take Lavasa, a "private city" being built in the lake-filled peaks of the Sahyadri mountain range some four hours southeast of Mumbai. India's cities will face major population growth in the coming decades, set to skyrocket from 340 million in 2008 to 590 million by 2030, according to McKinsey. To keep up with that demand, India would need to add some 700 million to 900 million square meters of floor space per year. By 2050, India will add 404 million urban dwellers; China will only add 292 million, according to a United Nations report. Current major cities are already straining under the pressure: Mumbai's population density is close to 53,600 people per square mile, more than double that of New York City.

Lavasa, being built by the Hindustan Construction Company, is the brainchild of CEO Ajit Gulabchand, who sought to make a completely new city with private funding on the banks of the Baji Pasalkar Reservoir. The five towns that will eventually make up Lavasa are being called independent India's first "hill stations," after the towns that nineteenth-century British colonizers built to escape the heat and the pace of life on the plains.

HOK, which had won a competition to design the new city's master plan, teamed up with what was then known as the Biomimicry Guild, a consulting firm cofounded by Janine Benyus, with the intent of making the hill city as environmentally friendly as possible. They took their cues from nature, examining the existing ecosystem—or at least, the ecosystem that used to exist, before so many slopes in the area were deforested by constant slash-and-burn agriculture. In the summer months when the monsoon rains sweep through, they wash soil down the bare hills, contributing to serious erosion. But those rushing rains only last a few months, and for the rest of the year, the area remains quite dry.

The team studied the local environment and realized that the barren hillsides had once been filled with deciduous forest, like much of the rest of the Western Ghats, which are known as one of the world's eight "hottest" biological hotspots, full of a rich diversity of plant and animal life. Such ecosystems provide a range of services for their inhabitants—they store rainwater, purify it, modify the climate, and mitigate extreme weather events, and produce new soil and cycle nutrients, among others. There's no intention to these systems—ecosystems aren't some kind of full-service biological hotel. Just as the termite mounds described in chapter 5 are the result of termites following simple rules—sometimes in competition with each other—ecosystems are emergent systems, a product of many different animals pursuing their own goals. A monkey relieving itself doesn't consider that it's providing fertilizer for the tree it's sitting in; the earthworm that munches through dung doesn't ponder the fact that it just happens to be releasing nutrients back into soil. These are relationships that have arisen through time and are constantly in just a little bit of flux. Nonetheless, just as the animals that survive are the ones attuned to the pressures in their current environment, the ecosystems that have emerged over time, thanks to the changing relationships between different species, are the ones that have achieved some form of homeostasis—where conditions are relatively stable. The only regulators, however, are the living agents in the ecosystem itself, constantly competing with and working with each other using the limited number of resources at their disposal.

In an ideal world, perhaps, you could create a computer simulation that would treat each necessary component or function of a city like an agent, as Rupert Soar described, and you could release those agents and watch what kind of system emerged. Barring that, the next best way to incorporate those ecosystem services is to take the bird's-eye view, which is the space that the Biomimicry Guild (now Biomimicry 3.8) folks seem to inhabit, and see what functions different ecosystems actually fulfill and analyze how they do it. Those specific processes are attuned to the local environment—and so, the idea goes, creating man-made systems that emulate those processes will be the smartest way to do so and will be best for the surrounding environment.

During the Los Angeles conference, Biomimicry 3.8 biologist and design strategist Jamie Dwyer gave an example pulled from the deciduous forest canopy that used to exist at the site. They initially assumed that the forest must have stored just about every drop of water it possibly could, with the canopy capturing rain and allowing it to seep into the ground rather than escape down the hillsides. Presumably, any buildings they made could be designed to capture rainwater just as well.

"They thought the best thing they could do was collect all the water and use [it] and then put it back into the water table . . . but in looking at how this ecosystem should function, we learned that twenty to thirty percent of the rainfall should actually go back up into the sky through either evaporation or transpiration," Dwyer told the crowd. "And if it doesn't, it actually changes the climate of that area."

This caught the team by surprise—after all, it seemed like that was simply wasted water. But releasing that water turned out to be essential to the larger environment's health.

"If the monsoon rainstorm is coming and the air is dry, it loses its momentum and it runs out of steam and it doesn't go very far inland," Dwyer explained. As the monsoon hits the mountains it tends to release its water, depriving the lands farther east. But, she added, with that extra humidity in the air from the trees, "then the monsoon storm comes in further. So you can see how something like that, it's a complicated system, but all of these details need to be in place in order for the rain to keep coming that far inland."

HOK's plan, then, would ultimately involve planting on the order of a million trees to help cover some 70 percent of the deforested land, as well as designing roof tiles for buildings that were based on the banyan fig tree leaf, whose long, narrow tips channel rainwater and can aid in its collection. They designed buildings with water reservoirs that were akin to a tree's taproot and circulation, as well as a drainage system that mimicked the grooved dams of harvester anthills to manage and divert excess storm water during the rainy season. These are a few innovations touted by the Lavasa plan, which has earned HOK several awards, including from the American Society of Landscape Architects.

But Lavasa doesn't have an entirely happy ending—at least not yet. In 2010, shortly before it was to have its initial public offering, the Lavasa Corporation was hit with an order to halt construction by the national environmental protection minister. The order and the ensuing legal battles dragged the project down by at least three years, Gulabchand has said, and has helped to bring the Hindustan Construction Company's value down from 78.9 rupees in January 8, 2010, to 19.85 rupees on March 23, 2016. (That doesn't even reflect the full loss of value, given that the Indian rupee has lost value against the dollar.) Along with accusations of excessive hill cutting, the project has also been plagued with allegations of bribery, land grabs from the local villagers in the area (some of whom, according to news reports, had no idea the land was sold from under them until a bulldozer came to raze their home), and alleged financial scandals. Politicians and company representatives involved in the Lavasa project have strongly denied these allegations in the public record.

Lavasa was supposed to be completed by 2021; Dasve, the first of the five towns that make up Lavasa, is still only partly built. HOK's Fannin told me in late 2016 that Dasve phase one has been built, but did not immediately clarify what "phase one" entailed. News reporters who visited the town interviewed employees who worked in the sparsely populated area but said, when the time came to settle down, they would raise their families elsewhere. Time will tell, whenever Lavasa is finally finished, whether it will truly become the template for future cities. For now, it mostly seems to be what one *Guardian* report called a "quirky weekend getaway"—for those who can afford it.

That is not to say there aren't truly interesting and less fraught projects happening at the city scale. In Puget Sound, for example, a team of consultants is looking to incorporate materials into building façades and sidewalks that manage Seattle's prodigious amounts of rain the way that the nearby forests do. (Among the concepts: evaporating the right amount of water back into the air the way that trees do, which may sound familiar from the Lavasa project; and building curbs out of material that, like mushrooms, stores rain until it can evaporate later.) The idea is to reduce the amount of storm water that simply picks up pollution from the impermeable roadways and becomes runoff, putting stress on the city's infrastructure in the process. This project is still in planning stages; it will take years to see how well these ideas really work.

The future of sustainable living depends on us remaking cities to behave in line with natural processes on every level, from micro to macro—for example, by making the materials out of which they're built smarter and more environmentally friendly, by using resources within buildings more intelligently and by designing smarter infrastructure for cities as a whole. But that is an enormous step to take, and one fraught with bureaucratic peril. For better or for worse, some of the most important innovations in sustainability may not originate at the level of cities, but within companies. That may be why the government entities interested in nature-inspired design seem to be looking to foster that kind of innovation within companies first. Take the New York State Energy Research and Development Authority (NYSERDA), which recently wrapped up a five-year program that looked to encourage biomimicry within businesses.

One of the business owners to take advantage of the NYSERDA program was Bob Bechtold, who as a young engineer got bit with the environmental bug hard. Bechtold was an early adopter of renewable energy—he started his work machining tools in his barn in Webster, New York, in the late 1970s and installed his first wind turbine on his eighty-nine-acre farm in 1980. A geothermal heating system soon followed. Making his home as close to carbon-neutral became something of an obsession—and in the 1990s, his mind inevitably turned to ways he could apply that to his nascent company, Harbec Plastics Inc.

Bechtold has a gentle way of speaking and resembles a heavily mustached Dick Van Dyke; it doesn't surprise me that he managed to persuade his neighbors to let him erect a behemoth of a wind turbine, an eighty-foot-tall machine, inspiring gawkers to slow down as they drove by it. But, as he told me, the idea of building an environmentally friendly company was a hard sell to lenders and customers alike.

"I started out, and made an enormous mistake initially when I took it to business: I was using the same passion in my personal life to try to accomplish business deals," Bechtold recalled. "And I was inadvertently discrediting myself, initially, as a burnt out hippie or a Birkenstocker because I was talking about all of the personal reasons—you know, my children's future and their children's, and all the amazing potential of free energy and all these kinds of things. Well, I learned the hard way that that was not the way to do it."

Discouraged, Bechtold went underground for a few months, retreating to the farm to rethink the story he was telling potential business partners. The key wasn't emotional or environmental, he realized—it was economic.

By the late nineties and early aughts, he'd honed his pitch. "My talk to anyone that will listen is only about the economics of the different opportunities that I saw in using resources more efficiently," he said. It seemed to work: He set up the company's first wind turbine, a 250-kilowatt machine, in 2001—a $425,000 investment that paid itself off in eight years. A $2.3 million, 850-kilowatt turbine followed a decade later. Together with the combined energy-heating power plant built on site, these systems account for 50 to 60 percent of their total energy use. He also built a combined heat and energy plant, and buys carbon offsets from the grid in the form of renewable energy to make up for the rest.

"We began taking, at that time, responsibility for our energy, because we're a large energy user," Bechtold said of his plastics company. "And we wanted to try and do it—and we didn't even know the word 'sustainability' back then—but in a more sustainable way."

In the mid-2000s, the engineer attended a NYSERDA-sponsored workshop on biomimicry by Terrapin Bright Green, a consulting firm that focuses on environmentally friendly product design and architecture.

Bechtold was intrigued by the ideas in Benyus's book and presented by Terrapin—how the elephant's ears act as a radiator, how termites regulate the temperature in their mounds. (Termites don't really, as Scott Turner's research has shown in recent years, but it's an idea that has taken on a life of its own in biomimicry circles.) He left the workshop thinking about the ways he might be able to apply nature's designs to his own company. After all, he'd invested all of this time and effort in making most of the energy he used "clean"—but what if he could use natural solutions to make his process more efficient—and thus save time, energy, and money?

"So Bob's like, 'I've heard of this thing but I'm just an engineer from the old days and I have no idea how to do 'biomimicry,' can you show me that process?' " said Cas Smith, an engineer at Terrapin.

With the help of a NYSERDA grant, Bechtold and Terrapin began looking at ways to apply natural solutions to Harbec's process. The key is to look for problems, Smith said—there's no point in trying to apply a solution from nature if you don't have a problem. And while the engineers at Harbec didn't think they had problems, per se, they certainly had points in their injection molding process where they had to spend more energy than they'd like. Take, for example, the plastic-cooling process. The company makes highly specialized plastic parts by injecting hot liquid polymer into metal molds and then cooling it using cold water that runs in channels built into the mold's walls. The cold water absorbs the heat from the plastic and then carries it away. The process takes a few seconds, but when you're making a million or so parts for a customer, all that time adds up. And if you try to save time and take the part out before it's fully cooled, it will most likely end up deformed.

"They were talking in terms of plastic and pumping and dwell time," Smith said. "That's very industry-specific to them. So we kind of abstracted the problem out and changed the language and said, 'Okay, it's really just a thermal transfer problem.'"

This act of simplification and translation is the key to bioinspired thinking, Smith said: you have to show that problems that seem very specific are in reality outgrowths of a more general issue—one that nature has probably tackled in a diverse set of ways.

Armed with this insight, Smith and his colleagues delved into the scientific literature, looking at the ways that different organisms manage heat transfer. They brought several examples back to the team, including mammalian lungs, termite mounds, webs of veins (as you might find in an elephant ear), as well as the veins in leaves. Lungs and termite mounds were ruled out, because those processes require air, which is far less dense than water and so doesn't carry enough molecules at normal pressures to be a truly effective coolant.

Eventually, they settled on leaves, testing out whether the monocot or dicot configurations were better. In thin monocot leaves (such as blades of grass), the veins all run largely parallel to one another. In wide dicot leaves, they tend to form netlike, interconnected networks—which they decided would be better for their system because they provided a larger surface area on which to do the cooling.

The company builds the molds using a form of additive manufacturing, where they use a laser to precisely weld bits of metal powder together. That meant that changing the design of the channels in the molds might take a little extra time but ultimately didn't really cost the company anything extra in terms of materials. And of course, once that mold is built, it can be used over and over again, saving the company precious seconds with every plastic part that is cast and cooled.

Bechtold had already tried out standard and "conformal" water channels; the conformal channels closely hug the contours of the mold, and the best of those were found to be up to 15 percent better at cooling molds than the standard variety. But the leaflike molds they created were even better—cutting cooling time by 21 percent compared to standard. Not only did it save them energy, it also saved precious time. And for a plastics manufacturer competing with foreign companies, getting your product to your clients as fast as possible is what helps keep that competitive edge sharp.

But when I ask him if his customers have noticed a difference, he demurs.

"We have to be careful about talking too much to the customer and getting them scared or worried," he said. "In a case like this, you have to be careful . . . only because it's brand new."

If a client knows of your dedication to sustainable practices, they might fear that they're not getting the best price, for example—an outgrowth of the misconception that environmental friendliness impacts the bottom line.

"It's a delicate talk of introducing customers," he added, "when they can handle it."

I ask him if he's met a lot of like-minded companies over the years, or if it's mostly been a lonely journey.

"Pretty lonely," he said, "but I'm happy to say it's improving notably lately." More companies are setting up appointments to study their heat-and-power plant, or asking for more details behind the figures Harbec has published on their website and in white papers.

On some level it's surprising to me that an engineer whose job is to make plastic parts has such a deep commitment to sustainability. But to Bechtold, even the plastic parts he makes could potentially be sustainable. For starters, he dismisses the idea of biodegradable polymers, arguing that they undermine the durability, strength, and other characteristics that make high-quality plastics so useful.

"If you were going to go to a cabinetmaker and ask him to build you a fine piece of furniture and then on the way out the door ask him to please insert termites, it would be counterintuitive, illogical," he said.

To Bechtold, the problem with plastics isn't just where they end up, but where they come from. Plastics are made out of petroleum products, whether oil or gas. The engineer focuses instead on bio-origin polymers—plastics that are made out of corn starch or vegetable oils, or with the help of microbes, for example. He's even put out an open call to the community, offering to test and analyze their bio-origin plastics.

"Most molders, their attitude would be, unless the customer demands it, and unless it comes with all of the proven set-up parameters that are required, they won't touch it, they won't bring it into the building," Bechtold said. "We invite the world to send us their material. . . . We will take a bag of their material, process the whole thing into a special molded part that we built that shows all the characteristics that you might want to use in molding a part, and send the molded parts back to them along with the set-up parameters for doing that work."

Even in this mission, the engineer sees a competitive advantage.

"You could say maybe that's altruistic to waste your time developing parameters for the world," he said, "but it isn't, because we become the people that are known to be able to do it."

I ask him about the potential issues with these bioplastics that echo those with corn-based ethanol; i.e., that the supposedly environmentally friendly alternative to oil-based products actually puts a strain on the environment because it requires the use of agricultural crops that eats into the food supply and still creates a significant amount of greenhouse gas.

"There are always going to be naysayers," he says, before pointing out that while we may find a way to make plastics even more cleanly in the future, we don't have that technology yet. "You take advantage of what you have at the moment to move it to the next level," he added.

As for the plastics' end, Bechtold argues that it doesn't make sense to design a biodegradable plastic—not just because it might impact the plastic's quality, but also because it implies that this plastic will ultimately be thrown away. To the engineer, that's like throwing value away.

"Every bit of polymer that's in the landfill today can be reused—every bit," he says. "Every single bit of can be taken back to basically the crude oil."

The only reason we toss plastic so easily right now is because oil is still pretty cheap. It's less expensive (economically, if not environmentally) to just drill for more oil and make new plastic than it is to mine the existing plastics currently sitting in landfills. But as oil becomes scarcer (and presumably more expensive, although recent prices seem to be bucking that rule), that calculus will probably begin to change.

"It's already technically true, but eventually it'll be finally true, that the landfills are full of our resources of the future,"

Bechtold says. "Including all of the plastic waste that people have been throwing away."

Part of looking to nature for inspiration, as Bechtold did, is looking to see how you can improve complex processes. Another, however, is recognizing that your building is part of the ecosystem in which it sits. But humans in cities rarely have a sense of what that really means. Take my home, Los Angeles, whose groundwater supports only 12 percent of its four million people. About 34 percent comes via the LA Aqueduct from Eastern Sierra Nevada, 45 percent comes from the Bay Delta, 8 percent comes from the Colorado River, and just 1 percent is recycled.

Of course, pretty much anyone who's watched or heard of the classic neo-noir film *Chinatown* has a sense of the shady way in which LA has acquired much of the water that allowed the city and the San Fernando Valley to grow well past its natural limits. William Mulholland, builder of the Los Angeles Aqueduct, and one-time mayor Frederick Eaton engaged in underhanded tactics in the early 1900s to build a 223-mile pipeline from the Owens Valley, which, ringed as it was by the Eastern Sierra mountains, was once called the Switzerland of California. While Eaton's cronies (including the owner of the *Los Angeles Times*, General Harrison Gray Otis and his son-in-law, Harry Chandler) plotted to buy up soon-to-be-valuable land in the San Fernando Valley, Mulholland deceived Owens Valley farmers by lowballing the amount of their water he planned to take. Otis published scare stories in his paper, warning of the droughts Angelenos would suffer if it did not build the aqueduct to quench its thirst.

"The *Times* quoted a disingenuous William Mulholland, who pointed to the city's dangerously low reservoir levels and publicly warned that Los Angeles was running out of water," Dennis

McDougal wrote in his book *Privileged Son: Otis Chandler and the Rise and Fall of the L.A. Times Dynasty*. "The *Times* did not report that Mulholland had been secretly instructing his workers to dump water from city reservoirs into the Pacific Ocean after midnight, when no one would notice."

The wastefulness of it all seems best summed up by Mulholland, who told a gathered crowd in 1913 as the Owens Valley water finally reached the San Fernando Reservoir: "There it is—take it."

In the end, due in large part to the rapaciousness of the San Fernando cabal, Mulholland decided to take essentially all the water from the Owens Valley, sucking it dry, and turning Owens Lake into a dusty salt flat. Those that live there today have to deal with violent dust storms that trigger asthma and other respiratory ailments.

Even with all that borrowed (stolen) water, the bill is still coming due. The city has been living in drought for the past five years; even as I write this, residents and reporters alike are asking why the long-promised deluge from El Niño seems to have bypassed a parched Southern California. To its credit, LADWP has managed to keep its water usage to approximately the same level as it was about four decades ago, when there were a million fewer people in the city; that's also to the credit of its citizens, who use 131 gallons per day in 2014 instead of 189 gallons in 1969.

Still, as Los Angeles's water woes continue—and even become more extreme in the face of climate change—it serves as a lesson for what happens when communities live beyond the resource limitations of their environment. Part of bioinspiration, then, is learning to live within those confines, to use resources in a more intelligent way.

Physical water scarcity affects about 1.2 billion people, according to the UN Department of Economic and Social Affairs, and 1.6

billion more people live in areas facing chronic water shortages (where there may be available water but no infrastructure to get it to people). As the human population continues to grow, it puts rising pressure on the water supply—especially since the rate of water use appears to be growing at twice the rate of population growth. And yet, the built environment isn't equipped to handle the water that reaches it in a way that replenishes and purifies the supply, the way that natural environments do. Concrete prevents the rain from sinking back into the earth to replenish the groundwater supply and clean it in the process. Instead, the storm water pools on those impervious surfaces, creating serious flooding issues in urban environments. (I can attest to this; even with a mild shower my neighborhood turns into a flood zone, three-foot-wide dirty rivers stretching out along the sidewalk curb.) A single inch of rain in Los Angeles can result in more than ten billion gallons of runoff, which makes its way into storm drains and washes out to the ocean. We are literally flushing a precious resource down the collective drain.

I wondered if cities were thinking about using their natural resources in more nature-driven ways and came across a pair of seminars moderated by Dominique Lueckenhoff, deputy director of the Environmental Protection Agency's Region 3 water protection division.

Both seminars involved bioinspired innovations: one focused on tapping into nature to address water issues, and one focused on learning from nature to design a better built environment.

Lueckenhoff's interest in these ideas goes back a few decades to when she was a very young environmental planner in Austin, Texas, and had the good fortune to meet Ian McHarg, a landscape architect who wrote the groundbreaking 1969 text, *Design with Nature*. In the book, McHarg urged his fellow planners to design in a way that worked with the surrounding natural ecology, rather than against it. Over lunch, McHarg told Lueckenhoff, with great excitement, about the Woodlands, a master-planned community about thirty miles north of Houston, built with the environment in mind. For example, the Woodlands' original drainage designs were not made of curbs with gutters leading to underground piping; they were open and on the surface, emulating the way storm water would flow had the forest still been in that spot. Later subdevelopments in the area switched to a conventional drainage system, and studies comparing the two showed that the original, nature-inspired system was actually much more effective at reducing runoff.

For the most part, McHarg's work was limited to designing in cooperation with the natural world, rather than focused on creating designs specifically inspired by the natural world. But those two ideas are symbiotic ones—they work best when they work together—and both require that you learn much more about the living systems you're interacting with. In that way, McHarg's career marked a major step forward in building towns and cities with far more sensitivity to the natural world.

Lueckenhoff's seminars were how I first learned about consultants like Terrapin Bright Green, a firm that helps their clients reach environmentally friendly solutions to their design and engineering problems. In both, Terrapin team members presented a variety of ways in which different companies were tapping into nature to take on the issues of water scarcity, including building bacterial cells that could treat wastewater while generating chemical energy, purifying and filtering potable water using low-energy methods, and building fog-harvesting mesh nets inspired by desert beetles. But one project that Terrapin worked on really struck me, which Terrapin partner Chris Garvin spoke about at the Los Angeles conference.

A company in midtown Manhattan had bought an entire city

block in Chelsea and had approached Terrapin several years ago with healthy but nebulous funding (perhaps somewhere on the order of $1 million a year), to devote to reducing their impact on the environment. They were interested in the idea of storm-water recapture—creating a building that could harvest any rain that fell on its surface. Of course, that's a much-needed technology, even in a place like New York, where water is currently plentiful but will grow scarcer over time due to climate change. (The folks at Terrapin would not tell me the name of the company, but it seems likely to me that the company was probably Google, based on references in the public record and on maps presented by officials that match the land the company sits on. Google did not respond to an emailed request for comment.)

This was a huge building, Terrapin partner Chris Garvin told the biomimicry conference attendees—holding roughly the same volume as the Empire State Building if it were turned on its side. Thus, any water reduction strategies they came up with had the potential to make a big impact on their footprint. The firm analyzed how much water could be saved by catching and utilizing the rain and snow that fell on the building, calculating that such modifications to the building would save some 5 million gallons of water per year. That sounds impressive, until you consider that when the firm totaled up the water used in the building's toilets, sinks, and kitchens, they were using on the order of 51 million gallons per year—roughly ten times the storm-water capture savings.

"It's very tangible," Terrapin's Cas Smith said of the storm-water capture strategy. "It's kind of sexy, if you do it in the right way and it's designed well and it looks neat. But if you're only at best able to capture 5 million but you use 100 million, then economically you're still importing 95 million. And then you have questionable effects from that imported water."

Questionable effects indeed—just ask the Owens Valley farmers in California. On top of that, the team analyzed how much water played a role in the company's energy needs. Like much of the United States, New York City draws power from thermoelectric power plants that heat up water from local watersheds and use it to drive steam turbines. Once that water is evaporated, it can travel anywhere else—which means that it's potentially lost to the local environment. By working backward from the energy bills, the consultants calculated that more than 500 million gallons were boiled away just to produce their electricity, bringing to total up to about 556 million gallons per year. And when the team analyzed the amount of water that was needed to make the food at the company's cafeterias—both of the meat and vegetarian varieties—the water load weighed in at a whopping 2–5 billion or so gallons, roughly ten times that of everything else.

Part of beginning to treat a company, a neighborhood, or an entire city like an ecosystem is not just to try and understand the natural environment's services in order to mimic them, as HOK did in Lavasa, but also to learn how to live within the true limits of the local ecology. Animals and plants do not, and cannot, directly import resources from hundreds or even thousands of miles away. (Migratory birds import themselves from one habitat to another; but even that is a relatively gradual process compared to today's human travel timescales, and involves behaviors and routes that have taken millions of years to emerge.)

All living things leave what humans might call a footprint. Some, like the globe-crossing arctic tern, can span hemispheres. Plants have reshaped rivers and sculpted coastlines. Apex predators as disparate as wolves and otters keep other species from overrunning an ecosystem, and help regulate carbon storage (by making sure the plants that store it aren't overgrazed). But "footprint" is a problematic term, because we think of it as separate from the foot that made it. Scott Turner has made a career of showing how the bodies of organisms do not truly end at their skins. Just look at the termite mounds he studies, which act as a lunglike extension of the termite colony itself. In his office, Geoff Spedding showed me a video of a tiny, translucent male copepod scuttling to follow the pheromone-filled eddy left by a female copepod. That trail through the water wasn't just some dead object separate from the female copepod, it was a part of her. The boundaries between organisms and their environment are transient, and the boundaries between different organisms and even different ecosystems are just as fluid.

Because humans can draw resources from far beyond the bounds of our local ecosystems, we expand our physiological footprint far beyond what we actually need. (As Muholland so aptly put it: "There it is—take it.") And we've created illusions for ourselves by building climate-controlled buildings that pipe their resources in from elsewhere, giving us the mistaken idea that the only footprint we have is the real estate we're standing on.

Part of Terrapin's work, Smith pointed out, is not just to make a company or a system work like the surrounding natural environment, which may frankly be infeasible given the amount of resources that cities suck up. Instead, the goal is to bring it, as much as possible, to within the limits of its ecological resources. The first step in doing so, as they did in this project, is to make that client aware of where all the resources they use are truly coming from.

So, to return to the billions of gallons of water used annually to make food for the Manhattan building's occupants: The foods that consume the most water to make are, largely, meats—especially beef. For a little perspective from the Water Footprint Network, it takes roughly 15,400 liters of water to produce a kilogram of beef, a kilogram of bread takes 1,608 liters, a kilogram of cabbage

280 liters per kilogram, and rice—which actually has to be grown in rice paddies, knee-deep in water—takes 1,670 liters of water per kilogram. To put it in dinner-plate perspective, a quarter-pound beef patty would cost you 1,750 liters of water, and feed one person. A 750-gram margherita pizza, in comparison, could feed three people for just 1,260 liters of water. The weak link in that food chain, then, is the meat.

"You can convince them to do rainwater catchment recycling—it'll probably cost a million dollars—and you're going to shrink down that 50 million gallons," Garvin told the audience. "You could do some really cool work replacing all the windows with these really cool windows that are more than 10 million dollars . . . or you could just do meatless Mondays."

Doing so, he added, would essentially cost nothing, and yet save more water than could possibly be saved by implementing all those more expensive, environmentally "sexy" solutions combined.

"It fundamentally changes the conversation with the client. You tend to look *kind of* smart when you bring this stuff up," he added with a little smile, "and it really gets them thinking differently about the systems with which they're interacting."

Up until now, all was as expected for the seasoned consultants, but here's where the story took a surprising turn.

"On another part of the day at that site, we were still talking about water and they just kind of mentioned, 'You know, our basement is really wet all the time and we have to pump out a lot of water; what do you think about that?'" Smith recalled. "And we thought, 'Huh, that's really interesting, your basement shouldn't be wet.'"

They weren't talking about a little damp in the cellar. The company had to use sump pumps to constantly bail out the water to keep it from drowning the basement, sending some 45 million gallons of water per year straight to the sewer. Keep in mind, Manhattan is essentially an island built of bedrock—so water should not be seeping up from the middle of it.

The consultants took a day to think about it, and turned to a unique resource at their disposal: the Mannahatta Project, which re-creates what the island looked like before 1609, when Europeans (namely, Henry Hudson) first arrived on scene. This effort from the Wildlife Conservation Society maps out every city block, allowing viewers to compare the island of the past with its present state. The project revealed that, around the area where the building now stands, there was plenty of flowing water. An underground stream, the team realized, could be the source. They went back and had the water tested, discovering that it was remarkably clean. So even as the company was trying to save millions of gallons of water, it was literally throwing out an annual 45 million gallons of usable water.

Armed with this knowledge, the consultants suggested that the building put the pumps to a new use—to gather the water for use around the building, from the cooling towers to landscape watering. For the moment, they're putting about 5–10 million gallons of that water to use annually, but they hope to use more as they scale up the infrastructure to utilize it.

"It showed the value of a historical perspective, that it can lend extremely valuable insights presently," Smith said.

This is what successful bioinspired design at the largest scales looks like. But in order for it to be truly successful, you have to apply it not at the ecosystems scale, but at the smaller process and even materials scale, in order to be truly productive. Nature's processes, after all, work on every scale, from nanometers to miles.

One company that's set the standard for employing these processes on small and large scales is Interface, Inc., which sent

Mikhail Davis to speak at the biomimicry conference. Davis first met Ray Anderson when he was an undergraduate at Stanford in 1999. A recovering biology major who grew up on a tiny commune full of therapists in Ojai, California, Davis had spent one solitary summer in college counting butterflies that were on the decline because their beach habitats had been cleared to make way for sunbathing humans. For seven out of the ten weeks, fog prevented the winged insects from flying, and the student soon realized that the biological sciences were perhaps not quite for him. Still, with his interest in biology and environmentalism, he figured he'd end up working for the government or for a nonprofit—which he did for a few years after graduation. But soon before he donned his cap and gown, he met, on the same day, Ray Anderson and Benyus, who were both speakers at a Stanford conference focused on business and the environment.

At the time, Davis was not expecting to be impressed by Anderson, CEO of Atlanta-based carpet-tile maker Interface, Inc. The Stanford student had a healthy dislike of corporations, Georgia companies even more so—one such company had deforested much of the land in his home county before leaving for cheaper opportunities in Mexico. But listening to this self-professed "radical industrialist" was a revelation.

"It was the first time I'd ever thought, 'Oh, I could do the work I want to do with a big company—you know, if there are guys like this,'" said Davis, who in 2011 became the company's director of restorative enterprise. "I had not been aware of the existence of his species before."

As a company that designs and manufactures modular carpeting, Interface seems, on its surface, as unlikely a candidate for bioinspired design and sustainable practices as Bechtold's plastic-making Harbec. Think of carpet tile, and you generally don't summon grand visions of the future; you think of bland office spaces and endless gray cubicle walls. Because that is, of course, the point of carpet tile: to create a practical, utilitarian, adjustable environment. Companies can avoid the expense of wall to wall; tiles can be more easily removed to replace stained sections or access wiring beneath the floors.

The company was founded in 1973 by Ray Anderson, a college football player at the Georgia Institute of Technology who graduated with a degree in industrial engineering in 1956 and eventually found his way into the carpet business. He discovered tile carpeting in England and decided to bring the concept to the United States, where he felt it would appeal to practical American sensibilities. By the 1990s, he'd grown Interface to a billion-dollar company, making it the dominant global player in modular carpeting.

"For twenty-one years I never gave a thought to what we were taking from the Earth, or doing to the Earth," he says in *The Corporation*, a 2003 Canadian documentary. That changed in 1994, when a customer in California asked a question that somehow trickled all the way back to Atlanta: what was Interface doing for the environment?

"The answer to that was, 'The what?' " said Davis. "This had not yet penetrated Georgia industrial circles."

Anderson, nonplussed, mulled the question. The company's research department, facing increasing questions from customers about environmental impact, had put together a task force to address the issue, and had asked the CEO to deliver a kickoff speech to the group at their inaugural meeting. Anderson agreed even though he had no idea what to say.

"I did not have an environmental vision," he said later. "I did not want to make that speech."

He was still desperately searching for inspiration when a book came across his desk: *The Ecology of Commerce*, by Paul Hawken.

In it, Hawken described the ways in which corporations were pillaging and polluting the planet—and the ways in which they could build a more "restorative" enterprise.

As he read, Anderson began to realize that he was a criminal—stealing from the future of the Earth, and the future of his great-great-grandchildren, to profit in the present. It was a painful epiphany—one he described repeatedly in the years that followed as "a spear in the chest."

Since then, Anderson decided that he would try to create a company that makes money without causing harm to the environment. The same year he read Hawken's book, he launched the Net Zero campaign—an effort to reduce the company's carbon footprint effectively to zero by 2020. In the modular carpet industry, that's not easy: it's extremely petroleum intensive, both in materials and production. And when they came across Benyus's book, they took to heart the idea of learning from nature to work with nature.

One long-standing problem with modular carpet is this: You can tell if a carpet square has been replaced, because it looks newer than the worn-out ones around it. That's the problem with having a uniform color, or even a uniform pattern (like a black-and-white checkerboard); the difference between new and old is still pretty clear. This was an aesthetic problem, but it was linked to larger financial ones: There's a lot of waste in the industry. Because the color-matching has to be so perfect, whole batches might be tossed out after manufacture because they're just a little bit off-shade. Because carpet tiles will inevitably require replacing, the company would recommend that people buy tons of extra tiles to store up for those occasions.

So for a biomimicry workshop, the company sent its employees on a little field trip, to forested areas outside of Atlanta.

"We send all of the carpet designers out into the woods,

at which point they thought we were totally crazy," Davis said. "It was very common in those early days—the folks back in Georgia were just like, 'Oh my God, what are they making us do now?' "

But as their shoes crunched on the dry leaves littering the forest floor, the designers began to notice a pattern, or rather a sort of non-pattern: reds and browns and golds of various shades, scattered throughout. There were no fixed grids or uniform colors—it was all random, but it all kind of went together, and it flowed in a way that was pleasing to the eye. This idea, of a design inspired by entropy, was completely foreign to a manufacturer like Interface, which deals in building and selling multiple identical widgets. Anything that deviates from those widgets is seen as off-quality, as waste. But the new patterns they designed, based on the random patterns of the forest floor, embraced the idea of entropy. These eye-pleasing designs weren't just good for business because they looked pretty; it also meant that they could be installed in any direction, replaced without needing an exact color match, and that customers would not have to stock up on extra tiles. Altogether, the change reduced installation waste by about 61 percent.

Interface also decided to tackle another sticky issue: The adhesives they use to glue the tiles down to the floor, which actually made it a pain to remove them and even risked damaging the floor. The company began to realize it was a real problem once they began recycling much of their customers' used carpet as part of their larger efforts toward Mission Zero.

"We'd be pulling it up and it was stuck to the floor and it was covered in glue, and it also had a lot of VOCs," Davis said, referring to volatile organic compounds that can worsen indoor air quality. "A lot of smelly carpet is actually the glue curing—it's not actually the carpet itself."

How could they get the carpet to stay in place without using these smelly, unhealthy, damaging glues? The designers took a look at how nature makes things stick. They were first interested in learning from the way that geckos use tiny hairs on the pads of their feet to cling to walls without a need for sticky adhesives—using instead the van der Waals forces that work to attract atoms to each other at nanometer scales. But the technology was not developed well enough for them to incorporate into their carpet backing. Then they realized: Why were they thinking of ways to fight gravity? Why not let gravity just do the work for them? Instead, they focused on ways to keep the carpet tiles connected to each other, eventually developing a plastic square the size of a Post-it that uses a resealable glue to hold the corners of four tiles together, eliminating the need for damaging, gassy adhesives. (The plastic squares, called TacTiles, were recyclable, too.) All in all, their environmental impact was 90 percent lower than traditional glues would be.

These and other innovations, many of them inspired by nature, have helped reduce the company's environmental impact as well. Since 1996, their per-unit greenhouse gas emissions have fallen 73 percent and their water intake per unit has dropped 87 percent. But it has also steadily improved upon their bottom line.

"We've made a lot of money off of biomimicry," Davis said.

Anderson exhorted other companies toward the moral path while never shying away from the role of prophet of doom, of the fact that American corporations were living in sin and needed to change their ways for the sake of their, and the world's, salvation. "In the future, people like me will go to jail," he told CEOs gathered at the U.S. Embassy in London in 1999, thus indicting, by association, those heads of business interests present.

Even as Anderson was traveling, speaking, and serving as an environmental evangelist, he worked to make his company not just a moral example but a financial one—a company that proved that the belief that the environment and the bottom line were in conflict was categorically false, and that you could indeed do well by doing good. Like Bechtold's Harbec, his company would serve as a model for to others.

Keep in mind, the carpet industry is extremely petroleum intensive—the nylon yarn and the backing is made out of fossil fuels and it requires a lot of energy to turn those fuels into useful polymers. So, the idea was, if a company like Interface can make these changes, any company can. Interface also has not shied away from throwing its weight around with its suppliers, favoring those who will meet its requirements.

"Carrot and stick, baby," Davis said, wryly. "You will lose our business if you do not come to play on sustainability. You will gain our business if you make this incremental [step] in the direction we want."

This leads to competitive sustainability, he explains. One yarn maker might offer to create a product with 5 percent recycled content. Another might up the ante, claiming 20 percent recycled content, but with a limited color palette. The next manufacturer might offer 25 percent with a fuller color palette. (The supplier who got to 100 percent recycled content has been getting a lot of their business, he added.)

Those kinds of pressures have resulted in programs like Net-Works, where the company that supplies Interface with its yarn doesn't pull petroleum from the ground, but instead pays impoverished fishermen for discarded nylon fishing nets that are often abandoned in the water, strangling reefs and killing fish, sharks, and dolphins. This way, the fishermen diversify their income streams even as they help clean their local environment, and the nets are reprocessed and used in Interface's carpets. It's a small but growing program: fishermen in the Philippines and Cameroon are collecting and selling some 3.5 metric tons of fishing nets every month.

It isn't always easy to stick to the vision, Davis said. He and his colleagues are constantly negotiating with those in the company who are directly responsible for the bottom line. A recent example: The nylon threads in their carpet yarn are made by taking white plastic, extruding it through what amounts to a spaghetti-maker with tiny holes and then dyeing that white plastic the desired shade. This isn't ideal, as it means the color eventually wears off and the product has to be replaced. A better way to do it would be to take already-pigmented plastic and turn that into yarn—built-in color cannot fade. That second method, however, is also more expensive.

"There's times when the head of procurement wouldn't speak to us for a few months," he says with a laugh.

Often (though not always), the more sustainable choice can end up making more sense financially, if you step back and take in the entire production process. In this case, the raw material might cost a little more, but it also meant they could do away with entire dye vats, get rid of the wastewater treatment system, and remove the yarn-drying system—eliminating a significant amount of operational expense.

"It's not like we don't have to push for change . . . it's just that it's easier for us," Davis says. "We still have to make the case—especially if it costs more money."

There's a fundamental weakness in the Interface model: Not every company has a Ray Anderson—exceedingly few do, in fact. But he lit that fire in others he met, including David Oakey, designer of the first "entropy" carpet tiles, and Mikhail Davis, skeptical former biologist. And while no other large companies seem to have dedicated themselves to the cause in the way that Interface has, many of them—including behemoths like Walmart and Unilever—are beginning to make the right kinds of purchasing decisions, thanks in part to the example Anderson set. And when they do, a whole host of companies up and down the supply chain follow.

"That was Ray's legacy," Davis said.

Anderson passed away after a nearly two-year fight with cancer in 2011. But the company has continued to follow his vision of reaching a truly sustainable enterprise by 2020. It's unclear if the company will be able to make it to zero-carbon and zero-water footprint by then; the third decade of the third millennium is approaching fast, and that goal may require some adjustment. Still, since the very beginning, the vision has never stopped at 2020, or even at "Mission Zero."

"I want to know what we'll need to do to . . . make Interface a restorative enterprise," Anderson told the task force in that 1994 kickoff speech. "To put back more than we take from the earth and to do good for the Earth, not just no harm. How do we leave the world better with every square yard of carpet we make and sell?"

Anderson wanted to get his company from a negative environmental impact to a net-zero and then finally to a net-positive impact; not simply to minimize the company's impact, but to make it an actual contributor to the natural environment. The idea is borne of an overt realization in recent decades that even in the so-called built environment of cities, ecosystems perform a number of services—purify and provide water, fertile soil, and sources of energy, for example—that urban dwellers and designers take for granted. Instead of ignoring and thwarting those services, such as filling in a wetland to make way for a parking lot, the new line of thinking finds that built environments should emulate those ecosystem services, providing as much natural value as the surrounding natural environment. I'm reminded of Mick Pearce, who wanted to make buildings like trees—buildings that capture and store rainwater, that provide a home for a diverse array of living things.

The company is already taking the first steps to this next stage in corporate sustainability. The plan, called "Factory as Forest," would look at the now-barren industrial park in New South Wales where one of Interface's plants sits and take inspiration from the natural environment that used to grow there. By exploring the nearby natural habitat, consultants with Biomimicry 3.8 have determined that the forest used to be a River-Flat Eucalypt Forest. The next step is to examine how the system functions and then use that as a benchmark that any factory in the area should reach. Among the services in a river-flat eucalypt forest? Pollination, carbon sequestration, water storage and purification, pollution detoxification, sediment retention, and fertilization, as well as resource recycling, to name a few. Once they analyze and measure how well it does each of those things, then they'll decide which of those services can also be incorporated into the factory building and premises.

"How are you going to accomplish something like pollination?" I ask.

"I don't completely know," Davis says. Some interventions, such as water retention, may involve incorporating rainwater capture systems into the building; they may burn the methane gas leaking from nearby landfills for power and figure out how to turn their carbon-dioxide emissions into artificial limestone, the way that sea creatures pull carbon out of ocean water to build their seashells.

But then again, he points out, they're essentially still in the planning stages. "It's a pilot . . . we're not supposed to know what we're doing."

My attendance at the biomimicry conference in Los Angeles was a well-timed accident—I'd arranged an interview with Terrapin's Chris Garvin in New York who (possibly in an attempt to escape my incessant questions) mentioned that he'd be headed to LA to speak at the meeting in two days' time. The community of self-described practitioners of biomimicry is small, and even more so in Los Angeles. Among the speakers is Ilaria Mazzoleni, who I had spoken to just weeks earlier about her work with Geoff Spedding on airplane design. Mazzoleni, who teaches at the Southern California Institute of Architecture, grew interested in drawing inspiration from nature after reading *Body Heat* by Mark Blumberg, which examined the ways that animals regulate their internal temperature. That, after all, is something that buildings are supposed to do, too: Maintain a comfortable interior temperature, keeping variation within narrow, acceptable limits. The more she thought about it, the more she realized that living bodies were the perfect inspiration for architects. Like the walls that define a building and protect it from the outside, a plant's surface or animal's skin protects its organs—but it also allows for sensing, chemical communication (pheromones), and exchanging materials with the outside environment, among other functions. In humans, skin also manufactures vitamin D from the sun; in plants, the surface harvests light to make sugar. Why aren't building envelopes, as architects call them, optimized for just as many functions?

It's an idea that feels familiar after watching Scott Turner and others work in Namibia, treating the walls of the termite mound not so much as a barrier between inside and outside, but instead as an interface between the two. Mazzoleni described a number of different riffs on that idea, pulled from concepts she and her students developed in workshops on nature and design (and which she later cowrote a book about). Among them: the polar bear, whose body structure she first analyzed before coming up with a design. A page on her site describes the bear's outer layers: the transparent hairs funnel UV light down to the bear's black skin; white wooly fur traps heat close to the body; and the bear's skin maximizes the absorption of solar radiation. She also took inspiration from what Turner might call the polar bear's extended organism—the den that a mother bear might dig to share with her cub. The architectural design this inspired was an oval room partly submerged in the snow that was angled to capture maximum solar energy and used giant glass tubes (mimicking the transparent polar bear hairs) protruding from the surface to help gather light.

Among the other speakers: Nicole Isle, a sustainability consultant at engineering firm Glumac; Colin Mangham, founder of Biomimicry LA; Lorraine Francis, director of hospitality interiors at Gensler; and Heather Joy Rosenberg with USGBC-LA, who's interested in ideas of resilience—particularly of note in a place like Southern California, where wildfires have always been part of the natural cycle. The keynote address was delivered by Amanda Sturgeon, CEO of the International Living Future Institute, a nongovernmental organization with the goal of fostering more sustainable buildings, communities, and products.

Sturgeon speaks of her grandfather, a former boxer in working-class England who built a rough-and-ready little greenhouse and filled it with sweet pea and other fragrant flowers when she was a child, kindling her love for nature. As she runs through several architectural projects, she focuses not on biomimicry, but on "biophilia"—a term popularized by famed biologist E. O. Wilson, who defined it as "the urge to affiliate with other forms of life." The buildings she features in her talk aren't exactly inspired by nature, but they invite nature in—incorporating trees into the roofless centers of buildings, or perhaps creating façades that are built into hillsides and covered with so much greenery that the edifice is almost invisible.

Certainly biophilia is a part of the story—if you love nature and want to be around it, you're more likely to listen to its lessons, after all. Still, it strikes me as such a different tone, and a way of thinking, compared to all the scientists and engineers I've spoken to for all the previous chapters. Many of the speakers dealing more directly in biomimicry also seem to echo this biophilic bent, speaking at length of childhood experiences with nature. Mazzoleni has gone even further, setting up a fellowship program where the selectees spend the summer in the tiny village in the Italian Alps where she's from, connecting with and drawing inspiration from the natural systems around them. A few describe epiphanies that are almost outwardly religious.

There are parts of the lectures where this starts to feel like a stretch to me, including what-would-Jesus-do questions like, "How would Nature design an office?" (My answer: it wouldn't.) Religions require stories, and stories occasionally overwhelm their facts.

Multiple speakers refer to "successful" examples of biomimetic design such as termite mounds and sharkskin, without mentioning that some of the science underlying those buildings or products remains either flawed or contested. If something worked, based on incomplete science, that is great. But when the science changes, when the picture grows more nuanced, it's important to keep returning to the science, rather than just to the story—even when speaking to the relatively uninitiated. If that's too much to swallow, then choose a different example, one where the biological model is still thought to be accurate. Because this sort of discussion implies, often mistakenly, that we know exactly how a given function works. If we keep marketing incomplete understanding of how nature works, then we risk stagnating the science, too. How many years passed before someone realized that we did not truly know how termite mounds work? Myths are powerful, and they can deter scientists from doing good research.

Take the polar bear inspiration: the idea that transparent polar bear hairs funnel ultraviolet light to the surface as if it were a fiber-optic wire was disproven in 1998, but the myth continues, perpetuated by nonscientists and scientists alike just enough to keep it alive.

Nature is not perfect; neither is our understanding of it. But plenty of popular narratives related to biomimicry appear to assume one or the other. And it's an issue that gives many of the scientists I spoke to pause. It may also be why some (though not all) of them seem to prefer the term "bioinspired design."

Take Robert Full, a UC Berkeley researcher whose lab I visited to learn more about his work with cockroaches and legged robots. He's guarded at first, asks me what my angle is, and then—as a sort of litmus test—asks me to read a quote by Benyus. I didn't write down the version he showed me on his computer, but it's a familiar and much-repeated quote that went something like this: "Ninety-nine percent of all species that existed on earth are extinct. The one percent here are the ones that work best."

"I think it's a decent idea; the only—"

"Completely wrong!" Full says, before I can finish.

"The only thing I would say," I continue, "is that nature is not necessarily perfect, not necessarily the best, it's just good enough, right now."

If you want an example of non-optimality in nature, just turn again to the polar bear. If an alien were to visit the Arctic's ever-shrinking ice and met these shaggy four-legged creatures, you can bet that they wouldn't imagine these animals to be the excellent swimmers they are. And if they were to design a swimming animal from scratch, it sure as heck would not have four legs.

One of my favorite examples is a blog called *WTF, Evolution,* in which an unnamed narrator asks Evolution why it chose to put a ridiculous black flap atop the comb duck's beak, or force hermaphroditic flatworms to "penis-fence," each attempting to spear the other with their phalluses without getting speared themselves. Evolution, ever enthusiastic about its creations, never really gives a sensible answer, probably because there isn't one. "Go home, evolution, you are drunk," the exasperated narrator often concludes. (That blog is now a book, subtitled *A Theory of Unintelligible Design.*)

I think my poorly worded but completed thought assuaged Full somewhat. Still, he continued, explaining an issue that's been voiced to me in many different ways.

"A lot of the popular things have this kind of thinking. And it's not how evolution works at all, which makes the whole design issue more challenging to try to explain and understand. So I think you had the right words, which this doesn't say," he says, pointing to the quote on the screen. "Of course there's no purpose in evolution; it's sufficiency and not optimality."

Therein lies the danger that many scientists see: Relying less on science and more on story leads ultimately to poorer design and applications, as well as the bane of overhype. When what's communicated sounds less like research and more like religion, it starts to make them uncomfortable. That said, some of the most ardent supporters of nature-friendly and nature-inspired solutions are those like Ray Anderson, who've had some kind of epiphanic experience.

Religion even came up during the conference when Kyle

Pickett, a managing principal at sustainability consulting practice Urban Fabrick in San Francisco, walked up to the mic during the Q&A session. Pickett grew up in the Pacific Northwest, raised in a "very conservative Christian home."

"Some of the conversations we're having with developers on a fundamental level is that 'man shall have dominion over the Earth,'" Pickett said. "And some of the conversations that I'm starting to have with these developers is kind of educating them to the extent that stewardship is really a fundamental quality within all of the world's major religions."

Pickett wanted to know if others had dealt with similar challenges; Mikhail Davis was the first to take the mic.

"We're a Georgia-based company so we have plenty of evangelical Christians—and a lot of them are our most zealous sustainability advocates. You have to leave room for people to understand this in their own ways. In the case of developers, it's very convenient to interpret sacred texts to support your own profit motives," he adds to laughter from the audience. "We have a preacher who runs one of our recycling lines who says, that that word 'dominion' is actually translated properly as 'stewardship.'

"That is actually a core and even a strength within our Georgia-based operation, these people who have incorporated our sustainability mission into their own personal spirituality," he continued. "So it doesn't have to go that way . . . I think you have to be open to people finding their own connection to biomimicry, sustainability. If you want to say these marvelous creatures were divinely created, great; let's mimic the best designer of all. I'm fine with that. I don't know the real answer."

There's something else Davis points out as he's talking with attendees after the conference wraps up. He mentions that they may be working on carpet filler with Brent Constantz, founder of Calera, Inc., which manufactures cement that actually pulls carbon dioxide out of the air—much in the way that coral pulls carbon out of the water to make carbonates for its shells. It's a potentially game-changing technology, given cement's ubiquity in extending the built environment upward and outward, and given that cement is one of the dirtiest industries, producing almost a ton of carbon dioxide for every ton of cement produced. Sometimes a successful technology, Davis says, is "so mature that we've forgotten that it's biomimicry."

Perhaps, when people no longer have to be "converted" into nature-inspired design thinking, when looking to biology for insight is so second-nature that it triggers zero double takes, when it applies to everything from materials to urban planning, perhaps that will be a sign that we're finally building a better future.

EPILOGUE

I started this book thinking I had a pretty good idea of exactly what I was going to say. I thought I understood the meaning of the terms biomimicry and bioinspiration. I thought I might even identify some good rules of thumb for identifying bioinspired designs. Looking back, even knowing that I thought these things with the confidence of ignorance, that surety boggles my brain.

Biologically inspired design sits at that perfect intersection between what P. W. Anderson called "intensive" research and "extensive" research. Intensive research means studying a particular subject in great detail until you truly understand its every aspect. Extensive research means taking a broad view of the research's implications and its applications, whether that means pulling its insights into a different field or building a device that takes advantage of the findings. Biologically inspired engineering lies at the nexus of these lines of thought—but that intersection is a constantly moving target, depending on the discipline of science you're in and the depth of knowledge already available to you.

There are so many researchers and so many lines of study I wish I had the time and space to tell you about. There are researchers like Robert Wood at Harvard University, who developed a tiny robotic bee that that is revealing the challenges of flight at such tiny scales. There's Graham Taylor at Oxford University, who puts cameras on hawks to film their flight paths and places flies in 3D movie theaters to watch how their bodies respond to changes in the simulated environment. His research could lead scientists to improve the software of future drones—to make it simpler, cheaper, and more resilient, much like the neurons in the fly brain governing flight. There are folks like Herbert Waite, who has spent decades analyzing the fundamental properties of mussel glues, enabling others to put those lessons to use in an adhesives market whose global value should exceed $50 billion by the 2020s. And there's Jeffrey Karp at Brigham & Women's Hospital, who has turned to all kinds of different creatures to inspire several different medical devices. That will all have to wait for another book.

As I reported and wrote these chapters, I kept encountering surprising crossovers, both in terms of characters and themes. In the chapter on swarm intelligence I talked a little bit about Marco Dorigo's work on robotic swarms—an idea I touched upon earlier in the chapter on non-wheeled robots. Ilaria Mazzoleni, who I first learned of from Geoff Spedding's work on birdlike planes, popped up at the biomimicry conference on the future of cities. People who are interested in bioinspired design often tend to be broad-minded, looking for common principles across widely disparate disciplines.

"It's exploding to a degree that is extraordinary," UC Berkeley scientist Robert Full told me when I visited him in early 2016. "I just had our board meeting for Bioinspiration and Biomimetics. And the doubling rate [of journals, conferences, and publications], which is a metric of the rate of interest in this field, is two to three years; the average for most other active fields is a dozen years. So it's extraordinary, actually, how fast things are changing. And so therefore it's important to try to really articulate what is real advancement in science and design—and what's not."

But there is a risk, as I mentioned in my last chapter, of going too broad and not deep enough to pull solid engineering lessons out of a natural subject. And when that happens, proponents risk setting the public (and potential funders) up for disappointment. It's a fear that looms large in the minds of scientists like Full.

"There is some concern that hype could move to an implosion with respect to acceptance by the businesses, funding agencies, and so we're really worried about that," he added.

Having seen the vast array of scientists and consultants at work—and the very different ideas they seem to have about what constitutes biomimicry or bioinspiration—I can't help but agree. It feels to me like a slightly fragile point in the field's evolution, a point at which people are looking for the applications before all of the basic science is there. Extensive science is always sexier than intensive science, which is difficult, incremental, and has driven many a Ph.D. out of academia. I meet those recovering doctorates everywhere: one used to be my boss; another I met while surfing in Santa Monica. Painful as it may be, that intensive research is absolutely necessary, because it provides a solid foundation on which to start thinking more broadly about the work's applications.

But matching natural systems with their potential applications is hard to do, even with consulting firms like Terrapin stepping in to fill that gap. That's why Georgia Tech computer scientist Ashok Goel is looking to create a system, powered by artificial intelligence, that could identify underlying principles in a natural system and, by analogy, match them to problems in need of a solution. The system is still in the works, but he hopes it could speed up the rate of bioinspired innovation to unprecedented rates.

There's another way to accelerate that process, Full points out: Get more diversity into science. And he's not just talking about race and ethnicity, although that's a part of it; he's also talking about socioeconomic status and background (did you grow up in a city or on the farm?), what your skills are—the formative experiences that give each person a unique point of view. Diversity, in all its forms, is essential for scientific innovation. Full has run bioinspired contests in the past and currently teaches a class on the subject and has been keeping an informal tally of the teams that earn top marks.

"I don't have data on this but I can tell you it was the most diverse teams—by far!" Full said. "The advantage it gives you for creativity is huge, if we can just support greater public support of education."

This makes sense to me. After all, it's easier to see the world differently, to make connections that no one has before, if you're looking at the world with different eyes.

ACKNOWLEDGMENTS

A book like this relies heavily on the generosity of many people—especially of those in it, who give of their time and energy and especially their patience. I regret that I cannot thank them all enough. I should first extend my heartfelt appreciation to Scott Turner and Rupert Soar for letting me crash their trip to Namibia on short notice and for calmly fielding my many (sometimes repetitive) questions along the way. It was the first chapter for which I did research, and it was a formative one. Thanks to Mick Pearce in Zimbabwe for being such a gracious host and for taking the time to show me his work, and to Lisa Margonelli for returning my forgotten camera. Deepest gratitude goes to Eric Nelson, without whom this book would not exist. Swati Pandey offered unending support, both of the chocolate and moral varieties. Thanks to Shakir, who always listened, and to my parents, who always asked.

NOTES

Numbers preceding citations indicate corresponding page within the text.

PROLOGUE

4. Benyus, Janine M. *Biomimicry: Innovation Inspired by Nature.* New York: Perennial, 2002. Print.

4. Fermanian Business & Economic Institute. "Global Biomimicry Efforts: An Economic Game Changer." *Economic Studies—San Diego Zoo.* San Diego Zoo Global, October 2010. Web. June 5, 2016.

1. FOOLING THE MIND'S EYE: WHAT SOLDIERS AND FASHION DESIGNERS CAN LEARN FROM THE CUTTLEFISH

12. Russell, Cary et al. "Warfighter Support: DOD Should Improve Development of Camouflage Uniforms and Enhance Collaboration Among the Services." *GAO-12-707.* U.S. Government Accountability Office, September 28, 2010. Web. June 5, 2016.

12. Cox, Matthew. "UCP fares poorly in Army camo test." *Military Times,* March 27, 2013. Web. June 5, 2016.

13. Rock, Kathryn et al. "Photosimulation Camouflage Detection Test." June 2009. Technical report. U.S. Army Natick Soldier Research, Development and Engineering Center, Natick, Massachusetts, 2016. Web. June 6, 2016.

13. Hepfinger, Lisa et al. "Soldier camouflage for Operation Enduring Freedom (OEF): Pattern-in-picture (PIP) technique for expedient human-in-the-loop camouflage assessment." *27th Army Science Conference,* Orlando, Florida, November 29–December 2, 2010. Conference Paper. U.S. Army Natick Soldier Research, Development and Engineering Center, Natick, Massachusetts, 2016. Web. June 6, 2016.

13. Campbell-Dollaghan, Kelsey. "The Strange, Sad Story of the Army's New Billion-Dollar Camo Pattern." *Gizmodo,* August 7, 2014. Web. June 5, 2016.

13. Campbell-Dollaghan, Kelsey. "The Army Is Finally Releasing Its New, Old Camo Design." *Gizmodo,* June 4, 2015. Web. June 5, 2016.

14. Deravi, Leila F. et al. "The structure–function relationships of a natural nanoscale photonic device in cuttlefish chromatophores." *Journal of the Royal Society Interface* 11.93 (2014). Web. June 6, 2016.

16. Darwin, Charles. *The Voyage of the Beagle: Journal of Researches into the Natural History and Geology of the Countries Visited During the Voyage of H.M.S. Beagle Round the World, Under the Command of Captain Fitz Roy, R.N.* New York: P. F. Collier & Son, 1909. Print.

22. Barbosa, Alexandra et al. "Cuttlefish use visual cues to determine arm postures for camouflage." *Proceedings of the Royal Society B,* May 11, 2011. Web. June 6, 2016.

27. Buresch, Kendra et al. "The use of background matching vs.

masquerade for camouflage in cuttlefish Sepia officinalis." *Vision Research* 51 (2011): 2362–68. Print.

38. Mäthger, Lydia M. et al. "Color blindness and contrast perception in cuttlefish (Sepia officinalis) determined by a visual sensorimotor assay." *Vision Research* 46.11 (2006): 1746–53. Print.
42. Yu, Cunjiang et al. "Adaptive optoelectronic camouflage systems with designs inspired by cephalopod skins." Proceedings of the National Academy of Sciences 111.36 (2014): 12998–13003. Print.

2. SOFT YET STRONG: HOW THE SEA CUCUMBER AND SQUID INSPIRE SURGICAL IMPLANTS

48. Coxworth, Ben. "Sea cucumbers could clean up fish farms—and then be eaten by humans." *Gizmag*, February 3, 2011. Web. June 5, 2016.
54. Capadona, Jeffrey R. et al. "Stimuli-responsive polymer nanocomposites inspired by the sea cucumber dermis." *Science* 319:5868 (2008): 1370–74. Print.
61. Miserez, Ali et al. "The transition from stiff to compliant materials in squid beaks." *Science* 319:5871 (2008): 1816–19. Print.
63. Fox, Justin et al. "Bioinspired water-enhanced mechanical gradient nanocomposite films that mimic the architecture and properties of the squid beak." *Journal of the American Chemical Society* 135.13 (2013): 5167–74. Print.
66. Prabhakar, Arati. Testimony to Subcommittee on Intelligence, Emerging Threats and Capabilities, U.S. House of Representatives. *Defense Advanced Research Projects Agency*. March 26, 2014. Web. June 5, 2016. bit.ly/1PemqVA.

66–67. Rudolph, Alan. "Nature's Way: The Muse." *Office of the Vice President for Research at CSU.* Wordpress, February 24, 2014. Web. June 5, 2016.

71. Khan, Amina. "For a 3-year-old boy, a risky operation may mean a chance to hear." *Los Angeles Times,* July 22, 2014. Web. June 5, 2016.

3. REINVENTING THE LEG: HOW ANIMALS ARE INSPIRING THE NEXT GENERATION OF SPACE EXPLORERS AND RESCUE ROBOTS

75. Calem, Robert E. "Mars Landing Is a Big Hit on the Web." *New York Times,* July 10, 1997. Web. June 6, 2016.

76. Khan, Amina. "Mars orbiter rediscovers long-lost Beagle 2 lander." *Los Angeles Times,* January 16, 2015. Web. June 6, 2016.

77. Khan, Amina. "Spirit's Mars mission comes to a close." *Los Angeles Times,* May 25, 2011. Web. June 6, 2016.

87. Pratt, Gill A. "Low Impedance Walking Robots." *Integrative and Comparative Biology* 42.1 (2002): 174–81. Print.

88. Glimcher, Paul. "René Descartes and the Birth of Neuroscience." *Decisions, Uncertainty, and the Brain: The Science of Neuroeconomics.* Cambridge, MA: MIT Press, 2004. Print.

90. Robinson, David W. et al. "Series Elastic Actuator Development for a Biomimetic Walking Robot." *1999 IEEE/ASME International Conference on Advance Intelligent Mechatronics,* September 19–22, 1999. Web. June 6, 2016.

93–95. Thakoor, Sarita. "Bio-Inspired Engineering of Exploration Systems." Jet Propulsion Laboratory. NASA Tech Briefs, May 2003. Web. June 6, 2016.

104. Marvi, Hamidreza et al. "Sidewinding with minimal slip: snake and robot ascent of sandy slopes." *Science* 346:6206 (2014): 224–29. Print.

107. Jayaram, Kaushik and Robert J. Full. "Cockroaches traverse crevices, crawl rapidly in confined spaces, and inspire a soft, legged robot." *Proceedings of the National Academy of Sciences*. 113.8 (2016): 950–57. Print.

4. HOW FLYING AND SWIMMING ANIMALS GO WITH THE FLOW

114. Huyssen, Joachim and Geoffrey, Spedding. "Should planes look like birds?" *63rd Annual Meeting of the APS Division of Fluid Dynamics*. Long Beach, CA, 21–23. November 2010. Web. June 6, 2016.

125. Fish, Frank and James Rohr. "Review of Dolphin Hydrodynamics and Swimming Performance." Technical Report 1801, SPAWAR Systems Center San Diego. August 1999. Web. June 6, 2016.

127. Hamner, W. M. Book review of Nekton. *Limnology and Oceanography* 24.6 (1979): 1173–75. Print.

129. Fish, Frank. "A porpoise for power." *Journal of Experimental Biology* 208.6 (2005): 977–78. Print.

129. Fish, Frank et al. "The Tubercles on Humpback Whales' Flippers: Application of Bio-Inspired Technology." *Integrative and Comparative Biology*. 51.1 (2011): 203–13. Print.

133. Kaplan, Karen. "Turning Point: John Dabiri." *Nature* 473.245 (2011). Web. June 6, 2016.

136. Gemmell, Brad J. et al. "Suction-based propulsion as a basis for efficient animal swimming," Nature *Communications* 6: 8790 (2015). Web. June 6, 2016.

5. BUILDING LIKE A TERMITE: WHAT THESE INSECTS CAN TEACH US ABOUT ARCHITECTURE (AND OTHER THINGS)

146. U.S. Department of Energy. *Buildings Energy Data Book: 1.1 Buildings Sector Energy Consumption*. March 2012. Web. June 6, 2016.

146. Campbell, Iain and Koben Calhoun. "Old Buildings Are U.S. Cities' Biggest Sustainability Challenge." *Harvard Business Review,* January 21, 2016. Web. June 6, 2016.

147. Goldstein, Eric A. "NRDC Survey: NYC Businesses Still Blasting Their Air Conditioners with Doors Open." *National Resources Defense Council,* August 26, 2015. Web. June 6, 2016.

147. Author uncredited. "Could the era of glass skyscrapers be over?" *BBC News Magazine,* May 27, 2014. Web. June 6, 2016.

152. McNeil, Donald G. Jr. "In Africa, Making Offices Out of an Anthill." *New York Times,* February 13, 1997. Web. June 6, 2016.

156. Turner, J. Scott and Rupert Soar. "Beyond biomimicry: What termites can tell us about realizing the living building." *First International Conference on Industrialized, Intelligent Construction (I3CON)*. Loughborough University, May 14–16, 2008. Web. June 6, 2016.

6. HIVE MIND: HOW ANTS' COLLECTIVE INTELLIGENCE MIGHT CHANGE THE NETWORKS WE BUILD

196. Tero, Atsushi et al. "Rules for Biologically Inspired Adaptive Network Design." *Science* 327.5964 (2010): 439–42. Print.

201. Gordon, Deborah. *Ants at Work: How an Insect Society Is Organized*. New York: Free Press, 1999. Print.

210. Prabhakar, Balaji. "The Regulation of Ant Colony Foraging Activity without Spatial Information." *PLoS Computational Biology* 8.8 (2012). Web. June 6, 2016.

216. U.S. Department of Agriculture. Cartographer. *Imported Fire Ant Quarantine*. Map. June 1, 2016. Web. September 14, 2016.

218. Buhs, Joshua Blu. *The Fire Ant Wars: Nature, Science, and Public Policy in Twentieth-century America*. Chicago: U of Chicago, 2004. Print.

218. Binder, David. "Jamie Whitten, Who Served 53 Years in House, Dies at 85." *New York Times*. September 10, 1995. Web. September 14, 2016.

218. Special to the New York Times. "Mississippi to Sell Ant Bait Despite Health Peril." *New York Times*. March 1, 1977. Web. September 14, 2016.

218. Sinclair, Ward. "Battle Against Fire Ants Heats Up Over Pesticides." *Washington Post*. October 13, 1979. Web. September 13, 2016.

218. Shapley, Deborah. "Mirex and the Fire Ant: Decline in Fortunes of 'Perfect' Pesticide." *Science* 172.3981 (1971): 358–60. Print.

218. Schoch, Deborah. "Aerial Spraying Won't Be Part of Fire Ant Fight." *Los Angeles Times*. March 12, 1999: B1 (Orange County edition). Print.

220. Anderson, P. W. "More Is Different." *Science* 177.4047 (1972) 393–96. Print.

224. Bonabeau, Eric et al. *Swarm Intelligence: From Natural to Artificial Systems*. New York: Oxford University Press, 1999. Print.

229. Spiegel, Alix. "So You Think You're Smarter Than A CIA Agent." *National Public Radio*. April 2, 2014. Web. June 6, 2016.

7. THE ARTIFICIAL LEAF: SEARCHING FOR A CLEAN FUEL TO POWER OUR WORLD

235. Editorial. "One and only Earth." *Nature Geoscience* 5.81 (2012). Web. June 6, 2016.

236. "Climate Change Impacts: Wildlife at Risk." *The Nature Conservancy*. Web. June 6, 2016.

237. Ciamician, Giacomo. "The Photochemistry of the Future." *Science* 36.926 (2012): 385–94. Print.

240. Fritts, Charles. "On a New Form of Selenium Photocell." *American Journal of Science* 26 (1883): 465–72. Print.

242, 247. Heller, Adam. "Conversion of sunlight into electrical power and photoassisted electrolysis of water in photoelectrochemical cells." *Accounts of Chemical Research* 14 (1981): 154–62. Print.

248. Fujishima, Akira and Kenichi Honda. "Electrochemical Photolysis of Water at a Semiconductor Electrode." *Nature* 238 (1972): 37–38. Print.

249. Khaselev, Oscar and John A. Turner. "A monolithic photovoltaic-photoelectrochemical device for hydrogen production via water splitting." *Science* 280.5362 (1998): 425–27. Print.

250. Turner, John A. "A Realizable Renewable Energy Future." *Science* 285.5428 (1999): 687–89. Print.

255. Kanan, Matthew and Daniel Nocera. "In Situ Formation of an Oxygen-Evolving Catalyst in Neutral Water Containing Phosphate and Co^{2+}." *Science* 321.5892 (2008) 1072–75. Print.

258. Liu, Chong et al. "Water splitting–biosynthetic system with CO_2 reduction efficiencies exceeding photosynthesis." *Science* 352.6290 (2016) 1210–13. Print.

265. Stoller-Conrad, Jessica. "Artificial Leaf Harnesses Sunlight

for Efficient Fuel Production." Pasadena: Caltech, August 27, 2015. Web. June 2, 2016.

265. Verlage, Erik et al. "A monolithically integrated, intrinsically safe, 10% efficient, solar-driven water-splitting system based on active, stable earth-abundant electrocatalysts in conjunction with tandem III–V light absorbers protected by amorphous TiO2 films." *Energy and Environmental Science* 8 (2015) 3166–72. Print.

271. Liu, Chong et al. "Nanowire–Bacteria Hybrids for Unassisted Solar Carbon Dioxide Fixation to Value-Added Chemicals." *Nano Letters* 15.5 (2015) 3634–39. Print.

271. Nichols, Eva et al. "Hybrid bioinorganic approach to solar-to-chemical conversion." *Proceedings of the National Academy of Sciences* 112.37 (2015) 11461–66. Print.

8. CITIES AS ECOSYSTEMS: BUILDING A MORE SUSTAINABLE SOCIETY

277. Kennard, Matt and Claire Provost. "Inside Lavasa, India's first entirely private city built from scratch." *Guardian*, November 19, 2015. Web. June 6, 2016.

280. Rossin, K. J. "Biomimicry: Nature's Design Process Versus the Designer's Process." *WIT Transactions on Ecology and the Environment* 138 (2010): 559–70. Web. June 6, 2016.

281. Press Trust of India. "Not involved in illegal land acquisitions in Lavasa: Sharad Pawar." *Economic Times*, October 8, 2012. Web. June 6, 2016.

284. Smith, Cas et al. "Tapping into Nature." *Terrapin Bright Green*, 2015. Web. June 6, 2016.

284. Lueckenhoff, Dominique. "Tapping into Nature: Bioinspired Innovation." *Faster . . . Cheaper . . . Greener Webcast Series: Connecting Natural and Built Systems for Economic*

Growth & Resiliency. USEPA Region 3 Water Protection Division. November 18, 2015. Web. June 6, 2016.

290. McDougal, Dennis. *Privileged Son: Otis Chandler and the Rise and Fall of the L.A. Times Dynasty*. Cambridge, MA: Perseus Publishing, 2001. Print.

291. Lueckenhoff, Dominique. "Biophilic Design for Human Health." *Faster . . . Cheaper . . . Greener Webcast Series: Connecting Natural and Built Systems for Economic Growth & Resiliency*. USEPA Region 3 Water Protection Division. November 5, 2015. Web. June 6, 2016.

300. Kinkead, Gwen. "In the Future, People Like Me Will Go to Jail: Ray Anderson Is on a Mission to Clean Up American Businesses—Starting with His Own. Can a Georgia Carpet Mogul Save the Planet?" *Fortune Magazine*, May 24, 1999. Web. June 6, 2016.

310. Koon, Daniel. "Is Polar Bear Hair Fiber Optic?" *Applied Optics* 37.15 (1998) 3198–200. Print.

310. Koon, Daniel. "Power of the Polar Myth." *New Scientist*, April 25, 1998. Web. June 6, 2016.

INDEX